HEART

Johannes Hinrich von Borstel is studying to be a cardiologist, and is also one of the best Science Slammers in Germany. He works as a paramedic in Marburg.

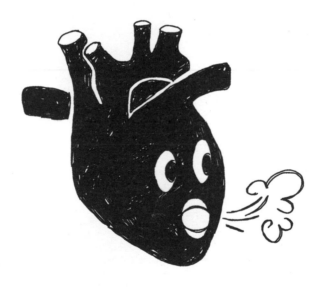

For Michi

HEART

the inside story of our body's most important organ

Johannes Hinrich von Borstel

Translated by David Shaw

SCRIBE
Melbourne · London

Scribe Publications
18–20 Edward St, Brunswick, Victoria 3056, Australia
2 John Street, Clerkenwell, London, WC1N 2ES, United Kingdom

Originally published in German as *Herzrasen kann man nicht mähen* by Ullstein in 2015
First published in English by Scribe in 2016

The advice provided in this book has been carefully checked by the author and the
publisher. It should not, however, be regarded as a substitute for competent medical
advice. Therefore, all information in this book is provided without any warranty
or guarantee on the part of the publisher or the author. Neither the author nor the
publisher or their representatives shall bear any liability whatsoever for personal
injury, property damage, or financial losses.

For the protection of the persons involved, some names, biographical details, and
locations have been changed and actions, events, and situations have been altered
in some places.

The moral right of the author has been asserted.

Typeset in Kepler Light 11.5/17pt by J&M Typesetting
Printed and bound in the UK by CIP Group (UK) Ltd, Croydon CR0 4YY

Scribe Publications is committed to the sustainable use of natural resources and
the use of paper products made responsibly from those resources.

9781925321807 (Australian edition)
9781925228809 (UK edition)
9781925307672 (e-book)

CiP entries for this title are available from the National Library of Australia
and the British Library

scribepublications.com.au
scribepublications.co.uk

Contents

Introduction

Everyone has a general idea of what a heart attack is: it's pretty bad news, health-wise. It causes chest pain and shortness of breath. Not infrequently, it can cause our heart, whose job it is to keep pumping blood through our arteries, to give up the ghost completely. Not good news at all. Our heart is, after all, the muscle that ensures even the most far-flung corners of our body, from the tops of our heads to the tips of our little toes, are kept constantly supplied with nutrients and, more importantly, oxygen-rich blood. This is, clearly, vital for our survival.

If someone were to interrupt the flow of blood from your heart to your brain even for just a few seconds, your body would react as if you had been hit over the head with a blunt instrument: you would lose consciousness, and whether your brain would be much more than blancmange afterwards is doubtful at best. This is because our brain doesn't handle oxygen deprivation very well at all. So, our heart beats — sometimes faster, sometimes slower, sometimes even seeming to stand still for a brief instant — an average of 100,000 times a day. Each time it contracts, it moves about 85 millilitres of blood, which makes approximately 8500 litres per day. We would need a tanker truck to transport that amount of liquid around with us. It's an impressive performance!

A heart attack was the reason I never got to meet my Grandpa Hinrich. He died more than a decade before I was born, after collapsing with pains in the chest and shortness of breath. Looking at the big black-and-white picture of him on my grandmother's living-room wall, I always used to wonder what it would have been like to meet him. Ironically, he looked so robust in photos! I could never understand how such a small thing could bring down such a fine figure of a man.

And so, from an early age, I began to devour all the textbooks and illustrated volumes I could lay my hands on that contained any information on the heart and how it can fail. My parents rewarded my interest by giving me more reading material, and I gradually began to develop a real fascination for all the processes that go on inside the human body. That was when I decided I wanted to work with nature and medicine when I grew up. I was determined to become a scientific researcher or perhaps a doctor (Plan B: street musician). And I wasn't content with just reading: I also collected everything from mouse skeletons to tortoise shells — anything that could help me gain a better understanding of the body.

When I was 15, I decided to make good use of the school holidays, put my books to one side, and apply for work experience at a veterinary clinic. Nervously, I dialled their number. I heard the phone ringing at the other end of the line. Four rings, five rings. With each second, I became increasingly anxious. Seven rings. Just as I had convinced myself that no one was going to answer, someone did pick up the phone. A woman's voice spoke in a business-like monotone.

'He-hello ... ?' I stammered. 'Is this the veterinary clinic?'

'Yes. How can I help you?'

I mustered all my confidence and replied: 'My name is Johannes von Borstel. I'm looking for some work experience during the holidays and ...'

The voice interrupted me, 'What year are you in at school?'

'I've just turned 15 and I'm in Year Nine.'

There was a heavy sigh at the other end of the line. 'Let me tell you straight out, there's not much chance of you doing work experience here. Sometimes, if we have an emergency situation, we might have to cut a dog right open without so much as a by your leave. You're too young to be watching that kind of thing.'

Too young? Surely not. Too squeamish? Possibly. That was precisely what I had to find out. It was the very thing I wanted to experience firsthand, to gain an insight into what happens beneath the skin, and to see with my own eyes all the goings on inside us mammals. How could I come by such an opportunity? I had no choice but to take the bull by the horns: I sent out more applications, including one to the emergency unit of my local hospital.

Two days later, the letter I had been eagerly awaiting arrived. A positive answer! And — I could hardly believe my luck — in the emergency department! At the time, I had no idea of the significance that piece of paper would have for my life. It was nothing less than my entrance ticket to a future more exciting than anything I had yet experienced.

The night before the first day of my work experience, I couldn't sleep. My head was buzzing with thoughts. Images of frantic emergency procedures, demigods in white coats fearlessly defeating every kind of disease, gaping wounds gushing with blood, and me in the midst of it all. I was wracked with nerves. What kind of medical cases would turn up the next day? What would I be expected to do? What would happen if I made a mistake? Might I make such a serious blunder that someone could actually die? And would it be my fault? I had no idea of the procedures on an emergency ward. My only preparation was a first-aid course I'd done.

'JOHANNES!!! FOR GOD'S SAKE! GET IN HERE NOW! HOW COULD YOU BE SO CARELESS?!' The voice boomed across the entire emergency ward.

Oh no, I thought. *I've really messed up. And on my first day, too.* Following the direction of the voice, I hurried across the ward and into the room I figured the ominous words had emanated from, to be confronted with a tragic *nature morte*. A doctor and an assistant stood before me, snorting with rage and glaring at me accusingly. Succumbing to the unstoppable force of gravity, drops of liquid were dripping onto the floor, where they formed a very conspicuous puddle.

'YOU'VE REALLY MESSED THIS UP! THERE'S NOTHING

4

MORE WE CAN DO. THAT'S IT NOW!'

I nodded, guilt-stricken, and looked away in shame. I had overreached myself. Then came the staccato orders from the doctor: 'Clean up this mess. The boss will be here any minute. We can't let him see this. He won't like it at all!' The assistant nodded in agreement and left the room. I pulled on some gloves, grabbed a roll of kitchen paper, and tore off a few sheets to soak up the accident. When the roll was finished and there was still no end to the deluge in sight, I threw in a towel for good measure.

I was just about to throw the rather pungent bundle in the bin when the senior consultant suddenly loomed up before me. 'Johannes? Have you put some coffee on?' He grinned, eyeing the dripping bundle in my hands.

'In 15 minutes ...' I stammered. 'I'll need to put a fresh pot on.'

The first mistake of my medical career: incorrectly filling the coffee machine, transforming it into a coffee-spouting gargoyle. Disastrous! It was the only coffee machine on the entire ward.

Well, that's a great start, I thought. *What can I say to the people in the staff room to turn this situation around?*

'Well, you'll just have to take your breaks without a cup of coffee,' I piped up, smiling hopefully at the assembled company. 'It could be worse, and it's healthier for you, too.' After all, this was a hospital. They should agree with my reasoning.

So, what did I learn on my first day? The simplest way to turn even the friendliest of hospital departments into a baying mob is to deprive them of their coffee fix. And acting like a pompous know-it-all compounded my mistake. No wonder, then, that I instantly rose from the position of intern to that of Public Enemy No. 1. To make amends, I baked them a marble cake.

The fact that I never actually had a serious mishap with a patient during my work experience was mainly down to the gradual, well-prepared way I was introduced to the jobs I was allowed to take on. As it turned out, I was not expected right away to treat gaping wounds, stem blood-spurting arteries, or deal with any other serious medical emergencies. Before I was allowed to join in with any such activities, I had to complete a very intensive program of learning and, most importantly, watching.

Shadowing the senior consultant on his rounds, learning bandaging techniques, practising the taking of blood pressure and pulse rates on the staff, entering data into the computer, and assisting in the treatment of minor to moderate wounds — this was my day-to-day experience as an intern. Additionally, the senior consultant would give me a short lesson at the end of each day, when he would explain in some detail the patient care and treatment strategies during the shift. He had a talent for explaining even the most complicated things in such a way that I could understand them, even without a medical degree.

Soon, I was learning to sew wounds. Well, okay, I started

on bananas. But, importantly, I learned that wounds are not always necessarily bloody. And perhaps most importantly of all, I came to understand the close connection between sensitive patient care and effective medical treatment. The senior consultant had the ability to recognise when patients were unhappy and to bring a smile back to their faces. He mentored me, far beyond the realm of medical issues.

With great patience, he explained the structure of the human body, from the skin down to the internal organs. And that's when I once again encountered my great (medical) love: the heart. Full of awe, I listened to him explain the muscles of the heart and its four-chambered structure. I heard tales of his time on call as an emergency doctor, of heart attacks, and of how to treat diseases of the heart. And the more I learned, the more my admiration grew for this fist-sized bundle of energy nestling in our chest. From that point on, I knew there could be no other — the heart had stolen my heart.

This book will take you on a journey to the heart. I begin by looking at the development and growth of the heart, and find out what that has to do with the theatre, loops, and bunny ears. I will also show you that our vascular system behaves like a network of highways — sharing all their features, from damaged roads to traffic congestion. You will see the sophisticated design of our heart, and how the processes of our atria and ventricles can run out of control. You will also learn what happens to our ticker when we smoke like a chimney, pay regular visits to McDonald's, and enjoy more than the occasional drop of the hard stuff. And I explain why emergency medicine has nothing to do with the esoteric arts but it's still necessary to be able to read coffee grounds.

Moving on, I will describe what diseases weaken our heart and pass on a few tips about healthy eating for the heart. I will then investigate whether the Easter Bunny would have a healthier heart if it were a vegan, why medieval physicians liked to quaff their patients' urine, and why Bucks Fizz isn't the only deadly quartet.

Next, we'll go on holiday together, but it will be a white-knuckle ride. The scene of the crime: the ventricles of the heart — since, as I will reveal, many a young holiday-maker's heart is less rested after their holiday than before. I'll clarify what exactly determines a healthy heart rhythm, what factors can influence it, and what we can do when that rhythm goes wrong. And I will take a look at the most drastic measure to get a heart beating again: resuscitation.

That's only required when one's heart stops beating — and to make sure this doesn't happen to you, I will prescribe a little something to prevent it: sex, which strengthens and supports the body and its defensive army, the immune system. I will zoom in on the tiny little warriors that make up our defence forces, and explain why Churchill's recipe for a long life ('No sports!') might not be the best advice. In passing, I will take you on a tour through our blood and its components, and have a look at blood pressure.

Finally, I will show that even our state of mind and

butterflies in the stomach can influence our heart. Is it possible to die of a broken heart? Whether it is or not, we should never underestimate the powers of self-healing. But modern medicine also has quite an inventory of tools to help repair a damaged heart, from replacing worn parts to installing a completely new engine.

These are the stations on our journey to the heart — each more fascinating than the last. And now it's time for that journey to begin!

The Loop in the Heart

How our heart develops, how it is structured,
and how its transport routes work

The Longest Theatre Play in the World

Ba-boom, ba-boom, ba-boom, ba-boom, ba-boom. The sound of a beating heart, powerfully performing its life-preserving service day after day. It beats without a break, no matter whether we're asleep or awake. It's already beating on the first day of our lives and continues until we draw our last breath. But what happens to our faithful ticker in the time in between, that is, during our lifetime? The answer is actually not very complicated.

I'm a passionate theatre-goer, and it occurs to me that the experience of a heart over its average 80-year existence is like a classical drama with five acts. The first act is the introduction. From the beginning of the second act, the action begins to rise. It reaches its climax in the middle of the drama, in act three. From that point on, all begins to go tragically downhill. After the fourth act, when everything moves from bad to worse, the fifth act ends with the inevitable tragedy, the curtain comes down, and the play is over.

But enough of this talk: the scene is now set for a real drama of the heart.

Act One: the unborn heart

In the theatre, plays usually begin by presenting the characters in the first act. So, allow me to introduce you. Very soon after an egg cell is fertilised, which is the point that marks the start of the rather complicated process of embryo development, the foundations are laid for the construction of a functioning heart. A rather unprepossessing collection of cells assembles, called the cardiogenic plate.* It forms two strands, which then develop into tubes.

At the same time, the pericardium, or heart sac, forms, and the heart continues its development inside this. The pericardium will later continue to envelop the adult heart. Inside the pericardium, the two parallel tubes now merge to create a larger one, called the tubular heart. It begins to move and eventually to curve in shape. Although it bears little resemblance to a rollercoaster or a display of aerobatic prowess, this process is called cardiac looping.

This isn't the end of the heart's development by far. Next, our heart grows ears — although not ones it can hear with. Like those fluffy bunny ears that are so popular at hen's nights, they only look similar to the real thing. Scientists are still unsure about the precise function of these heart-ears, which are in fact nothing more than appendages to the heart's atria. What doctors do know, though, is that they are responsible for the release of a hormone that will later stimulate urinary excretion. Our heart not only pumps blood

---

*   The word 'cardiogenic' comes from the Ancient Greek words *kardia*, meaning 'heart', and *genesis*, meaning 'origin' or 'creation'.

around our bodies, it also helps us to pee.

By this stage, almost a month has gone by since the egg cell was fertilised, and the embryonic heart can now be divided into recognisable sections that will become the chambers known as the atria (where blood enters the heart) and the ventricles (where blood is expelled). Precursors to the cardiac valves form, as do the early stages of the septum, or dividing wall between the right and left side of the heart. However, that wall does not form a complete partition in the embryonic heart, and will not fully close until a few days after birth.

In fact, there is an oval hole between the right and left atria, called the foramen ovale. Blood flows through this aperture from the right atrium into the left, and then on around the embryo's body. Why is that? The reason is simple: embryos are not yet able to breathe independently, so it would make little sense to invest in the laborious process of pumping blood through the embryo's lungs. This short-cut is all it takes to avoid that.

What eventually results from all this development is muscly on the outside and hollow on the inside (and thus could be said to bear a resemblance to a certain former governor of California).

Act Two: the newborn heart

The heart of a newborn baby is quite different from that of an adult. About the size of a walnut, it works much more

14

quickly. It beats up to 150 times a minute — even at rest: baby doesn't have to have been doing any sport. That's about twice as fast as the normal adult heart rate. The reason for this is simply that a newborn's heart is still very small and it pumps only a small amount of blood with each contraction. However, now that the heart is working entirely on its own, the foramen ovale closes during the first few days of life. With that connection blocked, the right side of the heart now pumps blood into the pulmonary circulation system of the lungs,* and the left side pumps blood round the rest of the newborn baby's body.

In the theatre, this is the stage when the first signs of conflict usually appear. The same is true of the heart. If something has gone seriously wrong with the development process of the heart, this is when it will become known, if it hasn't already. Although prenatal diagnostic techniques are now very advanced in the developed world, they are still not perfect, unfortunately. When doctors listen to an abnormal infant heart, they will often be able to diagnose a heart defect based on the sounds they hear.

The most common of these is what doctors call a ventricular septal defect, when the wall dividing the heart's two ventricles has a hole in it.† In the most serious cases, a young life must begin with major heart surgery. It depends on the size of the opening. Minor defects can heal up by themselves without any medical intervention, and as long as

---

*   *Pulmo* is Latin for 'lung'.

†   See also, 'The Holey Heart', p. 270.

the newborn child appears to be vigorous and thriving, there is no immediate danger to the baby's life. The decisive factor is whether the infant's organs are receiving enough oxygen. If this is the case, then doctors, parents, and, most importantly, junior can breathe easy.

Act Three: the strong heart

The heart of a healthy 20-year-old human contracts somewhere between 60 and 80 times a minute. If it is well trained, it can beat quite significantly more slowly when its owner is at rest. And this bundle of muscle is practically bursting with energy. The best way to gain an idea of its internal structure is to cut it open and take a look. For me, as a student of medical anatomy, this was an extremely exciting experience. But it might not be everyone's idea of fun.

The human heart

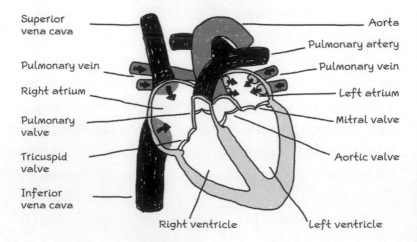

Superior vena cava	Aorta
Pulmonary vein	Pulmonary artery
Right atrium	Pulmonary vein
Pulmonary valve	Left atrium
Tricuspid valve	Mitral valve
Inferior vena cava	Aortic valve
Right ventricle	Left ventricle

16

Let's take a look at it from the point of view of a red blood cell, also known to scientists as an erythrocyte. It, and its many fellow red blood cells, gets its name from the red pigment haemoglobin, which it contains. Its main job is to transport oxygen from our lungs to the rest of our body, and, on return, to transport carbon dioxide back to our lungs.

Imagine you are an RBC (the slang term among medical types for red blood cell). You are transporting carbon dioxide — bonded to your haemoglobin — from one of the organs of the body, let's say the brain, through a blood vessel back towards the heart. So you must be in a vein, since that's the term for all the vessels that transport blood to the heart, while those that carry blood from the heart to the rest of the body are called arteries. After a few twists and turns, you eventually end up in the superior vena cava, a vessel that empties directly into the heart. And it is into the heart's right atrium that you are now swept, along with your cargo of carbon dioxide. From there, you pass into the right ventricle of the heart. Hurry now, don't dawdle, we have a mission to complete!

To get from the heart's right atrium to the right ventricle, you pass through an atrioventricular valve known to medics as the tricuspid valve (the Latin word *cuspis* means 'point' or 'tip'). Once you have left the right atrium via that valve, there is no going back — if you are in a healthy heart. All the heart's valves are unidirectional: they only let blood flow one way. This is a trusty means of making sure blood does not flow in the wrong direction, from the right ventricle back into the atrium. Thus, in a healthy heart, blood always only flows in one direction, and does not, for example, slosh back and forth between the ventricle and the atrium.

Continuing your journey, you leave the right ventricle via another valve — the pulmonary valve — heading towards the lungs. Having passed through that valve, you now find yourself in the pulmonary artery, the artery of the lung. This shows, by the way, that the much-quoted rule 'arteries transport oxygenated blood and veins transport deoxygenated blood' is in fact nonsense. After all, you're still carrying your cargo of carbon dioxide, making you 'deoxygenated', although you are currently floating through an artery, not a vein. Once more for clarity, the more accurate rule is: arteries carry blood away from the heart, veins towards it (although there are still some small exceptions to this rule, e.g. in connection with the liver*).

On arrival in the lungs, you complete the first part of your mission as an RBC by unloading your carbon dioxide and taking on a fresh cargo, this time of oxygen. With that freight on board, you now set out on a return journey through the pulmonary vein (!) back towards the heart. There, you and your many fellow erythrocytes flow into the left atrium and on, through a third valve, into the left ventricle, the last ventricle on your voyage. The valve between the left ventricle and the left atrium is known as the bicuspid† or mitral valve, so called because its shape reminded anatomists of the kind

---

* See p. 34 for more on the portal venous system.

† Meaning 'two-tipped'.

18

of bishop's hat known as a mitre.

The left ventricle is the bodybuilder among the chambers of the heart. It has by far the thickest muscle wall. This isn't surprising, since it needs to build up a great deal of pressure to keep our blood constantly flowing and to pump it to even the furthest reaches of our body. Now, on we travel, through a final valve, the aortic valve, and into the aorta, the body's main artery. This vessel describes a graceful curve around the heart, from which vessels branch off towards the head and the arms. It then continues into the abdomen, where it splits into ever-smaller branches to provide fresh blood to all our organs and tissues, right down to the tips of our toes.

We are now approaching the climax of our drama of the heart. Everything is working fine, the heart and vascular system seem to be indestructible. But things are about to take a tragic turn.

Act Four: the ailing heart

After just 25 years, the first 'deposits' begin to appear on the walls of the coronary arteries (arteries that supply the heart muscles themselves with blood). At this stage, it's not a big problem, but it lays the foundation for a very serious condition: arteriosclerosis, sometimes called 'hardening of the arteries'. It is the number-one cause of the world's most-common killers: heart attacks and strokes. Deposits of fatty plaques on the walls of the blood vessels will continue to build up, getting thicker and thicker and restricting the flow of blood until, in a worst-case scenario, a vessel eventually becomes completely blocked (like a water pipe with limescale).

When this happens to the coronary arteries, small or even larger sections of the heart muscle are left with an insufficient supply of nutrients and oxygen, and they begin to change. This is the infamous heart attack. Undersupplied areas of muscle transform into a kind of scar tissue that no longer contributes to the beating action of the heart. And, as we all know, a team is only as strong as its weakest member. The result is that the heart loses both strength and stamina.

At this point in a play, drama theorists speak of a 'delaying factor' in the plot, when the pace of the story slows down as the final denouement approaches. In the case of heart-attack patients, the role of delaying factor is played by medicine. To delay, or, even better, avert the oncoming catastrophe, doctors can prescribe medication, insert catheters (thin plastic tubes) into the coronary arteries, and try to alter the patient's life circumstances to take some pressure off the heart and thus minimise the risk of another heart attack.

Act Five: the (c)old heart

Chest pain. Irregular heartbeat. A listen to the chest with a stethoscope shows it is no longer beating with its regular ba-boom, ba-boom, ba-boom rhythm. Now, it sounds more like ba … boom, ba-ba-boom, boom, ba-boom. Difficulty breathing and weakness set in. After beating without a break

20

for almost a century, the heart is now significantly weaker. It's been through a lot. For some, it may be experiencing its second or third attack. It pumps ever less powerfully and, in a final act of valour, it gathers all its resources and tries to work faster. But in the end, all is in vain. The heart is no longer able to work properly; it twitches uncontrollably for a brief while and eventually becomes still. And then that's it: curtains.

This is the inescapable end to the drama. Predictable, but nonetheless tragic. Although all of our hearts will eventually stop, the time before this happens need not be dramatic. On the contrary, a hale and hearty life is more reminiscent of a comedy than a tragedy. The heart still ends up still, but at least its owner has laughed a lot and spent the time in a fulfilled way.

The good news is that anyone can take preventive measures to make sure the time when their heart stops beating comes as late as possible. And, at best, without cardiovascular disease ruining our existence before it does.

The first step towards this goal is keeping a sense of humour. Life can sometimes be an extremely serious business, but everything is easier when you're smiling. You could try laughter yoga. Or just search 'quadruplet babies laughing' on YouTube.

It's not only hypochondriacs who interpret trivial symptoms as the harbingers of death-bringing illness. No one is completely free of this crippling habit — not you, not me, not any of us. But the great thing is that, as a rule, human beings are basically healthy creatures. And that's true of the heart, too. When some part of our body feels strange, it's usually not due to a rare disease that will carry us off in a matter of hours, but due to something completely harmless. True to my favourite saying: 'If you hear hoof beats, think horses, not zebras.' So there is nothing standing in the way of personal happiness and physical health. But, still, I sometimes like to listen carefully to my heart.

Medics and Money before Midnight

I lie in bed and listen to my own heart hard at work. It's beating a little harder than normal because I swam a few lengths before going to bed. Looking at my alarm clock, I count 19 beats in 15 seconds. I do the maths: four times 19, or 19 times two times two. Or two times 38, equalling 76 beats a minute. I look down at my chest and watch it pulsating with each beat of my heart.

As a doctor-to-be, I have my stethoscope at hand. I listen to my chest. Ba-boom, ba-boom, ba-boom, ba-boom, ba-boom. I've just celebrated by 25th birthday. That means my heart has already beaten around 900 million times during my lifetime. Faithfully and dependably fulfilling its duty of keeping me alive. Thank you, dear heart, for doing that monotonous job for me.

But listening a little more carefully, it becomes clear that the work of the heart is not quite so monotonous after all. It does not simply go boom, boom, boom, boom like the bass from the speakers of your hi-fi system. In fact, a heartbeat consists of more than just the simple contraction of the entire organ. It is a precisely timed and coordinated operation of the muscles of the atria and the ventricles, as well as the opening and closing of the heart's valves.

First, the atria contract and press blood into the ventricles. This process can't normally be heard through a stethoscope.

A short time after, normally about 150 milliseconds, the ventricles contract, transporting the blood into the lungs and eventually the rest of the body. That contraction of the ventricle muscles is what causes the 'ba-'. The subsequent 'boom' isn't caused by the heart muscle itself, however, but by the closing of the semilunar valves of the aorta and the pulmonary artery. I place the stethoscope on another part of my chest. The sound changes. A little further up, and it changes again. I could happily spend hours just moving my stethoscope around.

What particularly fascinates me on this evening are the sounds made by the valves of my heart. They make sure that the blood travelling through our heart always moves in one direction and doesn't suddenly go into reverse gear. As we have seen, anatomists identify four valves, two of which are atrioventricular (the mitral and tricuspid valves) and two of which are semilunar.* Always alternating, they open and close, creating specific sounds that can be attributed to each valve. Medics distinguish between four heart sounds, although only two can be heard through a stethoscope.

The first of these heart sounds is caused by the contraction of the ventricular muscles. It is sometimes called S_1 by medics. The second, higher-pitched sound is shorter in duration than the first and is somewhat sharper and louder. Known as S_2, this sound is caused by the closing of the two semilunar valves. During inhalation, it can change and split into two components if the aortic valve closes a

---

* Meaning 'crescent-shaped'.

little earlier than the pulmonary valve.

As any parent or teacher will confirm, children and teenagers make more noise than adults — and the same is true of their hearts. The third and fourth heart sounds can't be heard through a stethoscope in a healthy adult, but they can sometimes be picked up in younger people. The third sound (S_3) is heard when the left ventricle of the heart is filled with blood. It is normal in pre-adulthood. When it occurs in older patients, it can be a sign of problems. More precisely, it can indicate a problem with the mitral valve between the left ventricle and the left atrium,* an abnormal increase in the size of the ventricular cavity,† or cardiac insufficiency (failure of the heart to work hard enough). And if the amount of blood remaining in the ventricle when it refills is too great, the new lot of blood will slosh against the old and this will also cause a sound.

The fourth sound (S_4) results from the contraction of the atria. If it's heard in adults, it can be an indication of high blood pressure, a thickening of the muscle of the ventricular wall, an obstruction in the left ventricular outflow tract, or — more rarely — a narrowing, or stenosis, of the aortic valve. It is generally followed immediately by the first heart sound.

Hearing all this with a stethoscope is a true art. Some doctors have such a finely tuned ear that they can hear not only the tiniest changes in the sounds of the heart, but

---

*   Also called 'mitral insufficiency', i.e. a failure of the mitral valve to close properly.

†   Ventricular dilatation.

also microtumours in the lungs. This involves placing the stethoscope on the chest and tapping in specific places. Theoretically, it's possible to identify such tumours by listening to the resulting echoes, but I have never managed such an impressive act. I suppose this is a case of constant 'practice makes perfect'.

Despite this, my stethoscope is always a great aid, for listening not only to the heart, but also to the rest of the body. I grew up in Germany's Harz Mountains, an area of the country that is very popular with motorcyclists in the summer. Serious accidents are common in the biking season and those horrific crashes often leave bikers with terrible injuries. On arriving at the scene of such an accident as part of an emergency medical team, I would begin by listening to the patient's lungs and abdomen. I did this because it's not uncommon to hear no sound of breathing on one side of the chest, even if the patient is still breathing.

The cause of this apparent contradiction is usually a collapsed lung (pneumothorax) on the side where no breathing sound can be heard, but it can also sometimes indicate an accumulation of blood in the chest cavity (haemothorax) or, in a worst-case scenario, a combination of both (haemopneumothorax). The sound made by tapping on the chest while listening through a stethoscope (doctors call this 'percussion') can allow a medic to distinguish between accumulated air and accumulated blood. An accumulation of air gives a sound reminiscent of beating on a drum, while an accumulation of fluid will dampen the sound, like striking a kettledrum filled with water. If the patient were able to sing and play the guitar along with all this percussion, they would almost be ready to take to the stage for a performance — if it weren't for the fact that they were in need of urgent medical treatment.

During a routine medical examination, the doctor will often listen to the abdomen using a stethoscope, to check the function of the intestines. After a motorbike accident, the medic will 'percuss' the abdomen to identify whether there is any internal bleeding or accumulation of fluids. As you can see, the stethoscope is a constant and valuable companion of medical practitioners; it is indispensable in many areas of treatment, but especially that of the heart.

However, like everything else, it has its limits. There are specialist cardiology stethoscopes with which you can almost hear the worms moving beneath the soil, but even they do not allow doctors to perceive everything — for example, the third and fourth heart sounds. For that, a special ultrasound examination of the heart is necessary (called an echocardiogram). It allows doctors to check the size of the heart, its atria and ventricles, the thickness of its walls, its overall mobility, its valves, and any defective blood flows. Often, a doctor will also be able to monitor for pathological changes to a patient's heart, including defects in the valves or constrictions in the coronary blood vessels.

While I was a medical student, I learned a mnemonic that has stayed with me ever since: '**All P**hysicians **T**ake **M**oney at **22.45**'. I'm sure all physicians are willing to take money at other times of the day, but what does this have to do with the heart? Well, this mnemonic helps young medics remember where to place their stethoscope when listening to the function of the heart valves.

Having memorised this little sentence, all a doctor has to remember is that the listening points are described from top right to bottom left. The time 22.45 describes the intercostal spaces (between the ribs), and the initial letters of the words in the sentence are the same as those of the names of the valves (**A**ortic, **P**ulmonic, **T**ricuspid, and **M**itral). Once you know this, you can listen very precisely to your own heart-valve sounds and any possible murmurs. However, interpreting these sounds is a complicated business and should be left to an experienced cardiologist, since recognising the subtle differences and changes is almost impossible without decades of practice.

There is a six-level scale for grading the loudness of heart sounds, ranging from 'difficult to hear even by expert listeners' via 'readily audible but with no palpable thrill' (medical word for tremor or vibration) to 'loudest intensity, audible even with the stethoscope raised above the chest'. In addition, doctors distinguish between different ways the sounds change over time, using criteria like crescendo or decrescendo, i.e. getting louder or getting softer; or diamond-shaped, which means getting louder then softer again; or a constant, unchanging intensity. The heart is an instrument that can play many kinds of music. Doctors use these distinctive features as the basis for diagnosing problems with

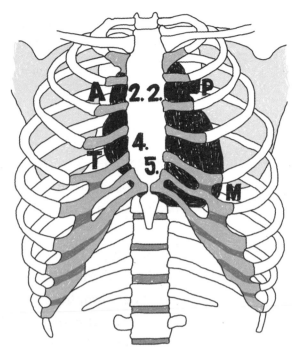

All Physicians Take Money at 22.45 — the stethoscope listening points

the valves of the heart, and prescribing the best treatment.

The way all the components of the heart work together is complex, but also absolutely fascinating. However, even the greatest, most powerful engine is useless if there are no roads for the vehicle to drive on. Our blood vessels are precisely those 'roads', without which our heart, as the central pump, would have no meaning. In the end, the heart's strength and stamina, as well as its intricate valve construction and conduction system, all serve one single purpose: to send our blood rushing at full throttle along those roads.

The Body's Autobahn

Our blood vessels transport blood and nutrients to the farthest reaches of our body. In fact, there are only a very few areas that are not permeated by them. Those include the corneas of our eyes, the enamel of our teeth, our hair, our fingernails, and the outermost layer of our skin. To transport all that blood, our body needs a proper system of pipes and ducts: our blood vessels. They are also the autobahns of our bodies. With the difference, however, that taken together, our arteries, veins, and capillaries (the finest branches of our blood vessels within tissues), are more than ten times longer than Germany's famous autobahn system — totalling around 150,000 kilometres.

Unlike the pipes that form the sewerage systems beneath our cities, blood vessels are very elastic. This is a good thing, because it allows the body to regulate the diameter of the blood vessels. It's what enables the body to provide certain organs and structures with a greater or lesser supply of blood according to whether they require more or fewer nutrients and oxygen molecules at any given time. When it comes down to it, this is no different to the engine of a car: the more you step on the accelerator, the more fuel is injected into the motor's cylinders.

When we are out jogging, our muscles need a better supply of blood to satisfy the increased need for oxygen. At the same time, our skin also receives more blood, so that it can release some of the increased heat into the environment via the cool, sweat-moistened surface of the skin. To make this increase in blood supply happen, our body reduces the amount of blood it provides to other parts of the body — for example, the gut. After all, digestion can always wait till later. A similar thing happens in our lungs: if a section of the lung registers a reduced oxygen supply, the vessels in that section will constrict. There is no point in sending blood to pick up oxygen where there is none to be had.

All this is possible because our arteries and veins have an elastic structure. The two types of blood vessel are similar, but there are certain differences between them. All have walls consisting of three layers, with the innermost layer made up of supporting connective tissue and what doctors call the endothelium. The endothelial cells line the inside of our blood vessels, serving as a barrier to protect the tissue

Structure of a blood vessel wall

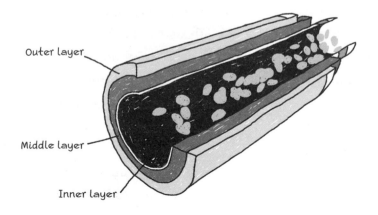

Outer layer

Middle layer

Inner layer

of the vessel wall, and they can play an active part in the regulation of the cardiovascular system. They are the interior decoration and the Anaglypta of our blood vessels, but they are also much more than that. For instance, they can release nitric oxide, which acts as a signal to the vessels surrounding the heart and those of the skeletal muscles to relax, allowing them to dilate. This happens during physical exercise and other times of exertion to supply more oxygen-rich blood to the muscles as they work.

The middle layer is muscular, or, more accurately, it consists of elastic fibres and smooth muscle cells that encircle the blood vessel. Here, the fibres of the autonomic nervous system — that is, the part of our nervous system that we don't consciously control — regulate the dilation or constriction of the vessel by tensing or relaxing the smooth muscle cells. Of course, the more dilated a vessel is, the more blood can flow through it.

The outermost layer of blood vessels is made up of connective tissue fibres, which anchor the vein or artery to the surrounding parts of the body. This layer houses those nerves that control the smooth muscles in the middle layer. But blood vessels also need oxygen themselves. This is provided by a network of tiny blood vessels, called the vasa vasorum. These 'blood vessels of the blood vessels' are also contained in the outer layer.

The arteries are like the sporty types within our bodies, while the veins are the couch potatoes. Their layered structures are basically the same, but the arteries are considerably more muscular. By the same token, veins have a larger internal diameter. These differences are due to the fact that the pressure inside our arteries is higher and they must be able to resist that pressure to avoid blowing up like wobbly, water-filled balloons.

Arteries can be divided into three types: elastic, muscular, and the smallest branches of the arterial system, the arterioles. The elastic arteries tend to be those closer to the heart, and the best-known artery of this type is the body's largest: the aorta, our main artery. It's about as thick as a garden hose. When the heart beats, the aorta dilates to accommodate a rush of extra blood, before contracting again to maintain internal pressure. Medics call this the Windkessel effect,* and it helps significantly in reducing fluctuations in the flow of blood to the peripheral areas of the body.

So, the arteries change their size by tensing or relaxing the muscles of their walls, and this regulates the amount of blood flowing to our muscles and organs. When they have almost reached their destination, the vessels branch out more and more to form arterioles. These get smaller and smaller until their walls are no longer made up of three layers, but just one layer of smooth capillary epithelial cells. At this level and smaller, scientists speak of capillaries. Every part of the body that has a blood supply contains a very extensive interwoven network of these tiny blood vessels, which can be

---

* From the German for 'air chamber', part of a water-pumping system.

so narrow that blood corpuscles* can only travel down them in single file, one behind the other.

The capillaries form the connection between the high-pressure arterial system and the low-pressure venous system. And since their walls are only one cell thick, oxygen can flow out of them and into the surrounding tissue much more easily than from other blood vessels. In fact, the endothelium is so porous that when tissue is infected and inflamed, white blood cells — which can be quite chubby little chappies — can leave the bloodstream through them. Eventually, the blood removes the carbon dioxide that has accrued in the body's cells and flows with it through venules and ever-larger veins, back to the heart.

Apart from a few exceptions, there is a clear division of labour between the arteries and the veins. In general, arteries transport oxygenated blood away from the heart, while veins take deoxygenated blood back to it. The exceptions to this rule are veins that transport blood directly from one organ to another without going via the heart — for instance, the portal venous system of the liver. This is the system that transports blood from the gut straight to the liver before it continues on to the heart. And that's practical, because some of the toxins we ingest along with our food are broken down in the liver before they can cause any damage in the rest of the body.

As we have seen, the pulmonary vein and artery are also exceptions to the general rule. The pulmonary artery, like

---

* Blood cells.

all its fellow arteries, leads away from the heart. However, it does not transport oxygenated blood, but rather blood that is on its way to the lungs to be enriched with oxygen. This blood then flows, packed full of oxygen, from the lungs, along the pulmonary vein, to the left atrium of the heart, where it is finally thrust out into the rest of our body via the left ventricle and the aorta, our main artery. This thrust is what we feel as our pulse.

The fact that our arteries are rarely to be found close to the surface of our bodies is a clever trick on the part of evolution, since a damaged artery bleeds heavily. Imagine the mess when you cut your finger while chopping carrots — and the risk of bleeding to death if you're unlucky. However, since our arteries are buried deep in our tissues, it takes more than a scratch to damage them.

Having survived a cut finger without bleeding to death, we need to ask how our blood gets from the tips of our fingers back to the heart. After all, it needs to return to the lungs to be recharged with oxygen. Before reaching the right atrium, it gathers in two large blood vessels, the superior and inferior vena cava. The superior vena cava receives blood from the upper body, the arms, and the head, while its 'inferior' counterpart gathers blood from the abdominal organs, the legs, and the torso.

But how does the blood in the veins of our lower legs manage the 130-or-so centimetre climb to the heart? This is only possible because our veins are equipped with small valves, positioned every few centimetres, that open for blood flowing towards the head but not the other way. Like the valves of the heart, they stop blood flowing back in the undesired direction. In addition, when we move, the muscles surrounding the veins do the rest of the work, pressing blood up towards the heart. This effect is appropriately known as the skeletal-muscle pump.

As we get older, more and more of our venous valves may stop working properly. When one valve gives up the ghost, the pressure is increased on the still-intact valve immediately below it, and the section of vein between them swells up. One unsightly result of this can be varicose veins, although they can also be caused by a general weakness of the connective tissue. This is often also the cause of another unpleasant vascular problem: haemorrhoids, which occur when the veins and arteries of the rectum swell and cause bleeding and itching round the back door.

However, it's not only the venous valves and the skeletal-muscle pump that are important in transporting blood back to the heart; this process is also aided by the body's respiratory pump. When blood has eventually made it back to the chest area, our breathing muscles help to transport it into the right atrium. This works because, during abdominal breathing, the pressure in the chest sinks as we suck air into our lungs, and that allows the inferior vena cava to take in blood more easily from the lower body. When we then exhale, the pressure on those vessels rises again, and the blood is literally squeezed into the right atrium of the heart.

As long as all these systems are working properly and all

parts of the body are well supplied with blood, there's usually nothing to worry about. Our cells get what they need and we carry on merrily with our lives. But it would be too good to be true if those systems were not prone to error. And, indeed, just like actual autobahns, the cardiovascular system is susceptible to congestion and, when push comes to shove, even gridlock.

Cardiac Congestion

Heart attacks and how they happen

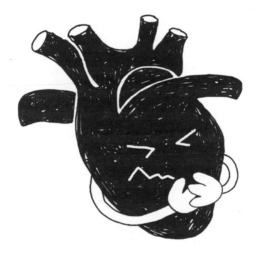

The First Time

It's a grey Saturday in autumn. The wind whips across the fields, and the rain pelts the tarmac in sheets. The streets are deserted except for the occasional car passing by. It has been more than a year since my work experience in the emergency department at the hospital. I'm 16 years old. Having completed the theoretical part of my training as a paramedic during weekends, I'm now ready for a three-week work-experience placement where I'll begin my practical training in an actual ambulance team.* I'm walking to the ambulance station to begin my first shift. The terrible weather does nothing to dampen my great expectations, nor does the fact that, in my excitement, I have not only forgotten to pick up my umbrella and put on my most-waterproof shoes, I've also left my packed breakfast at home. Luckily, it isn't a long walk.

Full of anticipation, I'm excited about my first ambulance call-out. What will it feel like? To be constantly on the go, flashing lights and blaring sirens, with total focus on the emergency at hand — illnesses, accidents, battles against the sheer forces of nature? I am ready to face any challenge! What I do not yet realise is that this is the day that will not only try my self-confidence to the limit, but also severely test my resolve to enter the medical profession at all.

After a few brief introductions in the station, I'm handed

---

* An unusual situation for someone my age, but I had great enthusiasm!

my uniform. It fits like a glove, and my chest swells with pride as I put it on. I am also given a handy little pager that will inform me of an emergency by beeping at me. Then I receive a tour of all the different machines and equipment in the ambulances.

As I'm chatting with my new colleague in the vehicle depot, the shift supervisor enters, with a slightly grumpy look on his face. 'Hello, Mr von Borstel, nice to see you here. I see you've been introduced to all the equipment and made some friends,' he remarks coolly.

'Yeah, yes, I have,' I stammer in reply. 'I'm very grateful for the opportunity to be here!'

He looks me calmly up and down, and a smile slowly spreads across his face. He tells me he has an important and highly responsible job for me. Ten minutes later, I find myself battling manfully against the sheer forces of nature for the first time. With a broom. In the driveway.

Is it a test? Some kind of initiation rite? Whatever: I don't care. Proudly sporting my hi-vis uniform, I do my leaf-sweeping duty in the rain. After an hour or so, I finish my fight against wind and weather, and withdraw to the staff room inside the station. In the staff room, there are couches, a TV, a small kitchen, and a bookshelf, from which I promptly choose some reading matter. Time passes, but there is not the slightest peep from my pager. My colleagues sit calmly with their pagers clipped to their belts, but I can't resist checking the battery level on mine every few minutes. Where are the emergencies? At lunchtime, we warm up some soup.

I wash the dishes. And nothing else happens.

It is unusual for absolutely nothing to happen during a 12-hour shift. Yet with two hours to go, it seems like we're heading for a shift without a single call-out.

Somewhat frustrated, I move downstairs to the vehicle depot and open the side door of an ambulance. I check the contents of every single drawer once more and try to commit to memory the way the emergency backpacks are organised.

And then, just as everyone has given up hope, it happens. I feel something vibrating on my belt and hear an insistent beeping. A call-out! My colleagues come rushing down the stairs, and a few seconds later we are tearing through the streets with blue lights flashing and sirens wailing. All we know is a name, an address, and the fact that the patient is having difficulty breathing.

Stefan, Sina, and I pull up outside a house. I grab my emergency backpack and the mobile oxygen cylinder, while Stefan takes the ECG machine.* We head straight for the front door of the house. I am buzzing with motivation; nothing can stop me now. Well, almost nothing. My mission comes to an abrupt and premature end when, in my boundless enthusiasm, I run headlong into the locked front door. Steady now! Ring the bell first. A light goes on.

'I'll be right with you,' we hear the voice of an elderly woman from inside. Through the glass door, we see the

---

* A device for recording the electrical activity of the heart in the form of an electrocardiogram (ECG). See also, 'If You Can See the Steeple, the Graveyard Isn't Far', p. 157.

silhouette of a human figure. She's hunched over and walks very slowly. 'I won't be a minute,' she says through the frosted glass. We wait. I'm charged like a live wire, but I can't help but be impressed by the sense of calm emanating from the woman behind the glass.

Finally, we hear the sound of the door being unbolted, and a lady with a snow-white perm opens it. She smiles. 'Do come in,' she says politely and ushers us inside.

'Did you call an ambulance?' asks Sina.

'Yes. My husband's in the living room. He's having trouble breathing again,' she sighs.

Weighed down by all my equipment, I trot after my colleagues down a dark hallway and into a barely brighter living room. The window blinds are half-lowered, and the flickering television screen is the only direct source of light. The furnishings in the room are old-fashioned — probably older than I am — but well cared for: a dark wood shelf unit, holding a few books and china plates; the television next to it; a coffee table with a brown-tiled top; and a couch, where a man aged around 75 sits, his face bright red. He is clearly struggling to breathe.

While I switch on the light, Stefan introduces us and immediately turns his attention to the patient. 'You called us because you can't breathe properly? When did this start?'

'I ...' he gasps with difficulty in response, 'I was just getting up from the couch, when ...' He pauses for breath. 'It was as if I was being strangled.'

In the background, I prepare the oxygen supply. I have

two ways to give the patient this life-saving gas: via a mask placed over his mouth and nose, or via a nasal cannula. The latter is a plastic tube attached at one end to the oxygen cylinder, splitting into two branches at the other end. Oxygen flows out of the branches, which are placed in the patient's nostrils. The amount of oxygen flow can be controlled by a valve on the cylinder.

I try hard to remember what I learned in training. Six litres per minute is the maximum amount to administer through a nasal cannula. Otherwise, the membranes inside the patient's nose are in danger of drying out. And in his condition, our patient has enough to contend with without that as well. After all, the oxygen is supposed to aid breathing, not make it harder. I could also opt for the mask. But then he will need at least six litres, or there's a danger the patient will not get enough of the oxygen. I waver. I might not give him enough oxygen through the cannula, but patients often find wearing a mask uncomfortable. After much consideration, I decide the patient will just have to deal with the discomfort of the mask.

Stefan ascertains the man's medical history and symptoms. 'Are you in pain, and if so, where does it hurt?'

'Here,' the man wheezes and points to his chest, on the left.

'Do you have any allergies?'

'No!'

'Do you take any medication regularly or have you taken any today?'

'No!'

'Do you have any other medical conditions?'

'Yes, diabetes.'

'Type 2?'

'Yes,' he coughs, 'type 2.'

'Do you take insulin?' asks my colleague.

'Oh, yes ... but just a little injection before each meal.'

Aha! This is something I was warned about in training, and now it's happening during my very first call-out. It is in fact extremely common for patients who take regular medication to deny it with full conviction when asked. I can't offer any explanation for this. It seems as if, for many people, taking medication regularly becomes routine, like brushing their teeth every morning. So they consider their pills, or even the contents of a syringe, no differently from the spoonful of sugar they take in their coffee or tea. It is certainly not deliberate deception on the patients' part — but in an emergency situation it can be deadly dangerous.

Stefan continues to question the patient about his medical history. 'Have you ever had difficulty breathing before, or have you been ill with anything other than a cold and your diabetes?'

'No, just diabetes!' answers the patient resolutely.

Yet suddenly, as if from nowhere, his wife joins the conversation. She has slowly but surely shuffled down the hallway into earshot. 'Tell them about your angina!' she shouts. 'Angiiiiinaaaa!'

With a slightly annoyed roll of his eyes, the elderly man

45

tells us he was diagnosed with angina pectoris* two years before, but no longer takes medication for it. He reports intermittent difficulty breathing, but says it always went away again and has never been this bad.

While Sina places the blood-pressure cuff on his arm, I offer him the oxygen mask, which he literally snatches from my hand and presses over his mouth and nose. I decide to begin with eight litres per minute. Using a pulse oximeter attached to his finger, I measure the oxygen saturation of the patient's blood. It seems pretty normal at the moment. But the man's blood pressure and heart rate are both high. This may be due to stress, or it could have a much more serious cause. Chest pain, breathing difficulties, and heart problems in the past — all the alarm bells are ringing.

My colleague takes an ECG reading while I prepare an infusion. As soon as the first lines appear on the ECG, our suspicions are confirmed: it's a heart attack!

Less than two minutes have passed since we arrived, and the patient's condition is rapidly deteriorating. He's having increasing difficulty breathing, and although I have turned the flow up to maximum, the oxygen saturation of his blood is plummeting. My colleagues do everything in their power to help him, while I feel rather at a loss. I follow my colleagues' instructions, preparing a needle and antiseptic for an intravenous catheter. As Stefan prepares to insert

---

* Latin for 'strangulation in the chest'; a temporary obstruction in the blood circulation of the heart, which is often associated with narrowing of the coronary arteries.

the needle, the man, now pale and blue-lipped, looks at me with fear in his eyes. His blood pressure is falling, his ECG is becoming ever more erratic, and the atmosphere is growing ever more sombre.

Sina speaks to him, trying to reassure him; the man never takes his eyes off me. His look screams, 'Help me!'

This is the worst feeling of my life so far. Inside my head, there's complete turmoil. What else can we do for him? Did my grandfather suffer like this? The man's gaze seems to pierce right through me. For a brief moment, I have the feeling that it's my own grandfather looking at me. Then, all of a sudden, the old man keels over to one side and loses consciousness. Before he can slide off the couch, Stefan catches him and lowers him carefully onto the carpet.

A quick check: breathing — yes; conscious — no. Place the patient in the recovery position and prepare to unblock the airways by suction if necessary; I remember my textbooks, and act accordingly. Get suction pump out of the rucksack, attach suction tube. A quick test and we're all set. If the man should start to vomit, I can quickly jump in with my pump.

The man's wife sits silently on a chair by the living-room door. We hear the howl of a siren coming from outside: the emergency doctor has been called and is on his way. Thank God for that! Sina asks the lady to go open the door. Just as she leaves the room, it happens: there's a piercing beeping noise, and the lines on the ECG begin to jump crazily. Ventricular fibrillation! A condition in which there is rapid and uncoordinated contraction and relaxation of the muscles of the ventricles of the heart, making it unable to pump blood.

Stefan immediately begins CPR (cardiopulmonary resuscitation) procedures, Sina prepares the defibrillator and I unpack the intubation equipment. At that moment, the emergency doctor enters the room. My colleague quickly fills him in on the situation, and then off we go: the man is defibrillated, which means we try to force his heart back into a normal rhythm with strong electric shocks. At the same time, we insert a tube into his trachea (windpipe) and ventilate him artificially, as well as giving him all sorts of drugs. For more than three hours, we try to keep him alive, but without success. This is my first emergency call-out, and it has ended in disaster.

When we return to the station later that evening, the nightshift team is already there, ready to take over. My colleagues hand the ambulance over to them as I dejectedly make my way home. I can't help wondering if I made some mistake, if there was anything more I could have done. Is this really the right job for me? Can I bear to watch people dying on a regular basis?

On arriving home, I study all the chapters on heart attacks in my collection of books for the umpteenth time, trying to find out where I might have gone wrong. This feeling of insecurity is new for me. It is some time before I realise: we didn't make any mistakes. For better or worse, I have to come to terms with the fact that even a trained paramedic can't save everyone.

Of Dragon Boats and (Oar) Strokes

A healthy human heart is about the size of a fist. Depending on body size and fitness level, it can weigh between 230 and 280 grams in an adult. It's made up mostly of heart-muscle cells, known to doctors as cardiomyocytes. There are two types of heart-muscle cell, and — a little bit like the staff on a hospital ward — there is a strict hierarchy between them.

The first type is the cells of the working heart muscles, which are responsible for making the heart beat by tensing and relaxing. They may be in the majority, but they can't escape the tyranny of the other, minority cell type: those of the heart's electrical-conduction system. Just like a pacemaker, these cells generate an electrical impulse and conduct it to the cardiomyocytes to cause the heart to beat. The two types of cell are like the drummer dictating the stroke and the rowers in a traditional Chinese dragon boat.

These two cell types differ not only in their function, but also in their appearance. The pacemakers are somewhat

larger than their colleagues, and have a 'pale and interesting' look about them. With impressive regularity, they make sure the heart beats at a constant rate — between 60 and 80 times a minute at rest. Provided, that is, that they are healthy and functioning as they should.

Unlike some of the other organs of our bodies, the heart has very limited regenerative capabilities. Compared to the liver, which can renew its cells exceptionally quickly, and even the lungs, which manage the same trick but at a much more leisurely pace, the heart is at the bottom of the regeneration league. Fewer than half the heart's cells are replaced over the course of an entire human life.

Despite this, the heart has all the cardiomyocytes it needs. The left ventricle alone is made up of an estimated six billion cells. If you were to look at each one of these cells through a microscope for half a second, you would have to spend almost 200 years at the eyepiece. That's without counting breaks for sleeping, eating, or satisfying other natural needs. Wow! That's a lot of cells! This naturally raises the question of where the heart gets all the energy it needs to pump some five to six litres of blood per minute, even when the body is at rest. The answer is simple: the heart is a believer in self-sufficiency.

Soon after the blood has exited the left ventricle to enter the body's circulatory system, it can take one of three possible routes. Most of it flows through the aorta towards our internal organs, arms, and legs. If it does so, it bypasses two exits just beyond the aortic valve, which lead to the right and left coronary arteries. Those vessels spread out into many smaller branches and supply the tissues of the heart with the nutrients they need.

At first glance, the pattern of those branches appears to be very similar from person to person, but a closer examination reveals that the details vary widely. Just like actual trees: a trunk in the middle, some branches, and a lot of leaves. It's not until you look more closely that you see each tree's particular characteristics, such as, for example, the pattern of the branches, the shape of the leaves, and the colour of the blossom.

In a left-dominant heart, the left coronary artery also supplies the posterior wall of the heart with oxygen and nutrients; in a right-dominant heart, this task is done by the right coronary artery. The most common type is the one in which both coronary arteries provide that supply in equal measure. This type is described by cardiologists as co-dominant.

Apart from forming branches, the coronary arteries can also form anastomoses. These are new connections created between blood vessels to make sure that effectively all the muscle tissue of the heart is constantly provided with the best possible supply of blood. Unfortunately, when one of the larger blood vessels becomes blocked in a heart attack, these anastomoses are almost never able to create a circulatory bypass as an alternative route to guarantee the continued supply of oxygen to the heart muscles.

When a heart attack occurs, the undersupplied tissues of

the heart begin to die. This can have varying consequences, depending on the location and size of the area supplied by the blocked artery. In the worst case, the heart simply stops beating immediately. If some of the rowers stop rowing, the boat may either start to spin round in circles or come to a complete halt. If the pacemaker stops beating its drum to dictate the rhythm, all the rowers will begin rowing for all they are worth, but out of synch with each other, so the boat still fails to move an inch. Sometimes, however, the reduced blood supply causes only slight irregularities in the rhythm of the heartbeat, and such very minor heart attacks often go unnoticed.

An arterial blockage — doctors call it an occlusion — leading to a reduced blood supply to the right side of the heart often causes the jugular veins of the neck to become engorged, since the blood flowing through those veins back to the heart can no longer be pumped into the pulmonary circulation system quickly enough by the right side of the heart. This leads to a traffic jam in the jugular. And no one likes a traffic jam.

An insufficient blood supply to the muscles of the left side of the heart, on the other hand, often leads to an accumulation of fluid in the lungs, known to doctors as pulmonary oedema. This is also caused by a traffic jam, but this time the blocked blood backs up in the pulmonary vein all the way to the tissue of the lungs. This builds up the pressure, causing fluid from the capillaries of the alveoli — the little air sacs in the lungs — to be pressed into the cavities that are normally filled with air, flooding them. This can be so conspicuous that a bubbling in the chest may be heard even without the aid of a stethoscope. In especially serious cases, the lungs can become so full of foam that the patient has to cough terribly hard to get rid of it. This can be a rather disgusting process, not only for the person concerned, but also for the emergency medical team treating the patient.

If no emergency doctor is on the scene at this point, the hands of a paramedic are basically tied. She can do little more than a trained first-aider. Of course, a paramedic can administer oxygen, but a first-aider can also simply open the window to help the patient breathe more easily. If the symptoms of such cardiac congestion become so serious that the patient's heart stops beating, anyone (not just a qualified medic) who discovers the unfortunate person should immediately begin resuscitation procedures. It would be good to have your last first-aid course fresh in your mind, but even less-than-correct resuscitation attempts are better than none at all.*

In addition to this, one thing is particularly important that has nothing to do with medical knowledge, machines, or electric shocks. It is providing for the patient's general comfort. This is so important because patients who are suffering a heart attack are often extremely frightened. The more anxious a person is, however, the more stressed he will be, and the faster his already-weakened heart will beat as a result. And that could be the final nail in his coffin, so

---

* See also, 'Quit Playing Games with My Heart', p. 168.

to speak. For this reason, it is crucial to create as pleasant an atmosphere as possible during the time until professional help arrives, and, again as far as possible, to radiate a sense of calm. When the patient feels he's in the care of someone who means him well, he will automatically feel better. If the patient is confronted with someone who is nervous and agitated, on the other hand, he will become increasingly anxious himself. Simply responding sympathetically to the patient's immediate needs is a great help. If the patient is cold, cover him with a blanket; if he has difficulty breathing, open a window. It has been proven that such simple actions can increase the patient's chances of survival, even in seemingly hopeless cases.

The same is true, of course, for stroke patients. That term is going to crop up several times, so let me explain it briefly here. A stroke* occurs in almost exactly the same way as a heart attack — only in a different organ. Our brain is permeated by a network of vessels that supply it with blood. This blood supply is important because our grey matter is made up of nerve cells that can only work if they receive sufficient oxygen via the blood. If one of the blood vessels in the brain ruptures, bursts, or becomes blocked, the area of the brain it supplies will no longer receive an adequate supply of blood and will die — unless the blockage is removed at once. By analogy with heart attacks, strokes are therefore

---

* Also known by many other names, including cerebral infarction, cerebrovascular accident, or cerebrovascular insult; formerly also apoplexy, palsy, or, in Latin, *apoplexia cerebri*.

also sometimes referred to as brain attacks.

Strokes can have very different consequences, depending on which blood vessel and which part of the brain are affected. Small blockages often go completely unnoticed, but if the area of the brain responsible for speech is undersupplied with blood, patients' speech may become slurred and indistinct, they may begin to say strange things, or they may lose the ability to say anything at all. When someone suffers such a cerebral infarction, time is of the essence. Within just a few hours, damage can become irreparable and permanent since, like the heart, the brain has very limited regenerative powers.

Of course, the best thing is to avoid any infarction — either cardiac or cerebral — in the first place. Treatment and care may be very good these days, but they are still unpleasant and dangerous. And it is, indeed, possible to reduce your risk of suffering a heart attack. Although, there are two factors we cannot influence: genetic predisposition and gender. Men are considerably more likely to suffer a heart attack than women. It's not until they have gone through menopause that women's risk increases, for which they have the change in their hormone balance to thank. But there are a whole range of factors that we can influence and that can increase our risk of suffering a heart attack immensely. If we avoid these risks, the danger is reduced. It can be that simple!

Russian Roulette with the Heart

The connection between smoking, drinking, and an unhealthy heart

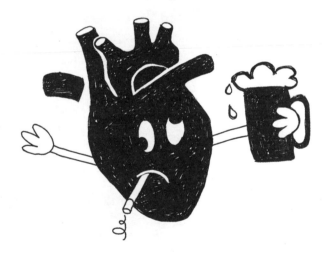

A Tar-Covered Road to the Heart

Why do we spend thousands on something that makes us stink, forces us to stand shivering in pub doorways in the winter cold, and causes us to die an early and unpleasant death? Why do we smoke, when it puts our cardiovascular system under extreme stress?

It's all the fault of dopamine, the pleasure and reward hormone in our brain. In our head, every cigarette is like a wonderful reward; it feels to a tobacco addict like a shot feels to a heroin junkie. This may be a simple answer, but it's no help to someone trying to give up smoking. Thankfully, however, we have public health warnings such as: 'Smoking causes ageing of the skin'. Boom! That really hits home! After reading that horrific warning for the first time, I felt awful, and my next cigarette was no pleasure at all. But was this any help to me? Of course not, because in such cases, addicts will simply increase the dose of the drug to force their brains to give them the reward that they crave, to make them feel better about the bad news.

I must admit that I set an extremely bad example myself when it comes to smoking. In anatomy lectures as a medical student, I saw smokers' lungs that were as black as a tarmacked road, and, working as a paramedic, I met people whose smoking habit had left them confined to a wheelchair or bedridden with severe chronic lung disease. But all this was not enough to put me off enjoying the occasional cigarette in the pub — despite the fact that smoking really is one of the few habits that brings absolutely no benefits; although those of a rather cynical frame of mind have argued that smoking benefits the state pension system by causing smokers to die an early, if medically expensive, death. Smokers are playing a game of Russian roulette, only without passing the revolver on, since the lungs are far from being the only organs affected by the more than 4000 toxins contained in tobacco smoke.

What exactly happens when we smoke, and how can it be so damaging to our bodies? Cancer! That's the first thought that springs to mind. Of all the substances we breathe in when we smoke, at least 40 are known to cause cancer. The biggest risk is that of developing the kind of tumour known in common parlance as lung cancer, which doctors refer to by the much more elegant-sounding term 'bronchogenic carcinoma'. This happens when some cells in the bronchial tubes begin to mutate and, with time, are no longer able to perform their normal function. The affected cells then try to make up for that loss by constantly reproducing. They divide and multiply incessantly, creating a tumour that eventually impairs the functioning of the lung. If these cells then also find their way to other organs via the bloodstream — when we say the tumour has 'spread' — our entire body is affected, and death is the final result. If we know all this, how can we be so stupid as to keep on smoking?

Nicotine is the guilty culprit. In small doses, it causes

a moderate release of adrenaline, the well-known stress hormone that stimulates us, suppresses hunger pangs, and increases alertness. What a wonderful natural drug! However, nicotine's main effect is to cause the brain to release dopamine, the pleasure hormone we met above. Nicotine also causes both heart rate and blood pressure to rise.

When I was 18, I did an experiment on myself to observe the constrictive effect of cigarette smoke on our blood vessels. A friend of mine happened to own a thermal-imaging camera, so I used it to film my own hand while smoking. Before I lit up my cigarette, the surface temperature of my skin was 32 degrees Celsius. But even after only one puff, that temperature fell to 30°C. And after I had smoked the entire cigarette, the temperature of my hand was fluctuating around the 28–29°C mark.

So nicotine and tobacco smoke are addictive substances whose effects on our body are not only long-term; they have some pretty immediate impacts. Causing our blood vessels to constrict can be very serious: if a smoker already has a narrowing of one of the coronary arteries, a single cigarette may be the straw that breaks the camel's back, causing the artery to close completely — and the smoker to keel over with a massive heart attack.

As well as nicotine, two other important substances contained in tobacco smoke are tar and carbon monoxide. The latter is a colourless, odourless gas that binds to the red blood cells, greatly reducing their ability to take up oxygen.

They are able to do this because erythrocytes (you remember them, I'm sure — the RBCs) take up carbon monoxide much more easily than oxygen. In the most serious cases of carbon-monoxide poisoning, the gas displaces oxygen from the blood corpuscles to such an extent that the resulting lack of oxygen is life-threatening. This explains why many people who want to kill themselves choose to do so by deliberately inhaling the exhaust fumes of a running car engine, which contain carbon monoxide.

Smokers will notice dark-coloured mucous when they cough. The substance responsible for this discolouration is tar, which blankets the cilia in our lungs. Cilia are the little hair-like structures whose job is to wave constantly back and forth (looking for all the world like a field of corn in the wind) to transport mucous and any inhaled foreign particles, such as dust, out of the lungs and into the nose and throat area. The smoke from just one cigarette leaves the cilia paralysed for several minutes. So frequent smoking throughout the day means rather a lot of unwanted material collects in the lungs, making our breathing apparatus more susceptible to infection and other causes of illness.

Nicotine and tobacco smoke also reduce the concentration of 'good' HDL cholesterol in the blood, and increase that of 'bad' LDL cholesterol.* Furthermore, they thicken the blood and damage the inner wall of the blood

---

* The section 'Should the Easter Bunny Go Vegan?', p. 118, deals extensively with cholesterol.

vessels. This is one of the main causes of arteriosclerosis,* a real roundhouse punch with negative consequences for our cardiovascular system. It's no wonder, then, that in Germany alone, 110–140,000 people die every year from the effects of smoking.[†]

If, to make matters even worse, these effects are combined with other risk factors for heart attacks, such as high blood pressure, increased blood-cholesterol levels, and life as a couch potato who's on first-name terms with every McDonald's employee in town, the risk of cardiovascular disease can reach enormous proportions. In addition, smoking is one of the main causes of peripheral arterial occlusive disease (PAOD), known to most non-medics as smoker's leg.

In this condition, the blood vessels of the legs become so damaged by deposits of fat and plaque that the patient has difficulty walking any distance. Sufferers have to take a break every few metres. Since this behaviour resembles that of a window-shopper, stopping regularly and often to admire window displays in town, it is sometimes known as 'window-shopping disease'. A walk round the shops is supposed to be a pleasant experience, but in serious cases of PAOD, the undersupplied tissue can die off, and must be surgically removed.

Furthermore, tobacco smoke also affects our immune

---

*   For more information, see 'A Tight Squeeze', p. 83.

†   In the US, the figure is 480,000; in Britain, it's 100,000; while in Australia, it's 15,000.

62

system, weakening it enormously. Men may smoke more and be more likely to suffer from vascular disease, statistically speaking, but that doesn't mean that smoking isn't also extremely dangerous for women. Especially women who take the birth-control pill. The contraceptive pill can increase the risk of the blood vessels becoming blocked by blood clots — a condition known as thrombosis. If a woman smokes while she's on the pill, she is combining two risk factors and thus hugely increasing the likelihood of negative medical consequences.

All this means that, when it comes to smoking, there can be only one sensible decision: to kick the habit as quickly as possible. Even long-time smokers will really be doing their bodies a favour by quitting, since many studies show that an ex-smoker's body will begin to regenerate soon after that last cigarette — slowly, admittedly, but continuously. Ah, that last cigarette; it really is the best one!

The first positive changes kick in just 20 minutes after the final puff. In that time, the ex-smoker's blood pressure will have returned to the level it was at before that last cigarette. The body's blood circulation and temperature begin to return to normal. After about half a day, the amount of carbon monoxide in the ex-smoker's blood will have sunk to a safe level. This allows the blood corpuscles to deliver pure oxygen to our cells once again. After just one day, therefore, giving up smoking will already have helped the heart. And that considerably reduces the risk of a heart attack.

Two days after quitting, we begin to smell better, and I

don't just mean our body odour, but also, more importantly, our *sense* of smell. At the same time, our sense of taste begins to regenerate, which increases our quality of life immensely. The taste of ripe Italian tomatoes can now really blow your socks off!

After two weeks, our lungs begin to perform better, and even after just a month, the cilia will have regained their pre-smoking functionality. This means much less coughing to expel mucous and dust from our breathing apparatus. The result is that we suck more fresh air into our lungs with each breath we draw.

Once six months have passed since that final cigarette, the risk of suffering a heart attack will have sunk by half. And if an ex-smoker holds out for another six months — so, a whole year in total — then his or her risk of dying due to the effects of smoking will be about half as high as it was immediately after that last cigarette. Then the worst is over, although the danger of slipping back into old ways remains present for years.

I know this from personal experience. In my final years at high school and during my paramedic's training, I no longer smoked. But when I went to Vienna to study, one single drag on a cigarette was all it took to put me back where I was at the time I stopped smoking. Unfortunately, we humans have an excellent memory when it comes to addictions. Even years after we quit, our body remembers how great the feeling was when we indulged our addiction. By contrast, the negative effects are soon forgotten. The simple fact is that our brains are greedy for dopamine.

Luckily, human beings also have the capacity to conquer their urges. For those who manage to quit permanently, their risk of suffering a heart attack will fall to the same level as a non-smoker within 15 years. It really is worth doing more than just thinking about quitting.

There's also no reason to be afraid of withdrawal symptoms after quitting smoking. It is true: lack of concentration, increased irritability, and often bouts of sweating and nausea make life difficult at first; but, when you think about it, these are positive signs! They show that the body is adjusting to the new, altered circumstances it now finds itself in. So the best thing to do is simply grin and bear it, and to make sure you never have to go through it again!

Bomb Shots for the Heart

There's nothing I like more than a night out drinking with friends. We usually go out and do something together before we finish the evening with a game of cards, a beer, and, for some of us, a smoke. Although it rarely turns out to be just one beer. However, what we rarely talk about, indeed, what we prefer not to think about, is what happens when the effects of smoking are compounded by those of consuming alcohol. This popular combination, a drink and a smoke, causes far more damage in our bodies than tobacco smoke alone.

A phenomenon we hear about a lot in the media is binge drinking — regularly consuming large amounts of alcohol in a single session. Researchers in Chicago carried out a study that showed that test subjects who were regular heavy drinkers do not profit from the much-propagated beneficial effects of alcohol on the circulatory system. Quite the opposite, in fact. The study recruited male volunteers between the ages of 18 and 25, some of whom were regular binge drinkers and others who were teetotallers and never normally drank alcohol at all. Over a period of two hours, the researchers gave each volunteer four to five standardised drinks, containing 13 grams of alcohol each. This is roughly equivalent to a 300 mL bottle of beer. The scientists then measured the diameter of the arteries in the volunteers' arms.

In the teetotallers, the measured arteries dilated both when they were given drugs to stimulate that dilation and when they weren't. By contrast, the amount of dilation in the arteries of the binge drinkers, who reported having got 'blind drunk' around six times a month in recent years, was much less pronounced.

We often read that drinking a glass or two of wine in the evening is good for our health, and in particular for our cardiovascular system. But to conclude from this that alcohol is some kind of preventive medicine is a fatal mistake. Alcohol is not a household remedy, but an addictive substance that increases the risk of myocardial disease, abnormal heart rhythms, and organ damage.* In addition, the liver also suffers massively when we drink alcohol to excess, and this, in turn, can directly affect the cardiovascular system.

Several studies have shown that alcohol consumption is the cause of about 40 per cent of the damage that occurs to the muscles of the heart. In a similar way to a heart attack, this means muscle tissue dies, again increasing the risk of life-threatening heart disease. Alcoholics also have a compromised immune system. Their army of defence cells no longer works as well as it did before their addiction, increasing the risk of infections, which can also spread to the heart.

Alcohol abuse affects the entire body. It damages the brain, and, in some males, it's not only their brain which shrinks, but also their testicles. The digestive system suffers

---

* See also, 'Holiday-heart Syndrome', p. 143.

when we drink to excess, too. Even casual drinkers will have noticed this at the toilet bowl the morning after enjoying one beer too many at a party. Whichever orifice it comes out of, the result is usually unpleasant and runny. Many extreme alcoholics barely eat anything at all, as their bodies can no longer tolerate food. But a healthy diet for the heart includes more than just hops, malt, and barley.

Despite all of this, drinking a glass of wine with dinner is of course not a crime for a healthy person with no pre-existing conditions, and a fit body can easily cope with a beer enjoyed in the company of friends.

Reading Coffee Grounds from the Ground

Nee-naw, nee-naw! We're stuck on the road. The ambulance siren is wailing, but we're not moving. The way ahead is blocked by a car engaged in a slow-motion, precision-parking manoeuvre. In his anger, my teammate Thomas hammers on the ambulance's horn, just to make it extra clear to the unruffled motorist that we urgently need to get past. Of course, the sound of the horn is completely drowned out by the ambulance's siren anyway.

'Great idea,' I drawl. 'If he hasn't noticed by now that we're practically blow-drying his hair with the siren, that pathetic whine from the horn is really going to wake him up ...'

We both get the giggles. It is a little macabre to make jokes when you're on your way to deal with an emergency. But the behaviour of motorists when they encounter an ambulance is often pretty bizarre. The first few times I encountered drivers who refused to let us through, cut off our right of way, or performed other absurd stunts, I would lose my temper. But I've learned to take it in my stride: after all, it's not good for the heart to get stressed out by such things.

At last! The meticulous parker has finished his manoeuvre and the way ahead is clear. Off we go at top speed. A little monitor on the dashboard tells us the address and some basic information about the patient. Male, 55 years of age. Suspected symptoms: gastrointestinal bleeding with vomiting. Working in the medical emergency services, you constantly deal with blood: blood samples taken from patients' arms; heart attacks caused by blood that can't reach the right place; or blood that is still inside the patient, but not where it should be, as in a stroke.

Most often, it's blood that's in the process of exiting the patient's body. The flow can be slow and oozing, but it can also trickle, or even shoot out in a thick jet. After all, some of our blood vessels have about the same diameter as a ball-point pen. If they're injured, the blood can spurt for metres. Bright-red blood indicates that the injured vessel is an artery. If the blood is darker and more bluish in colour, it usually comes from a vein.

For injuries that involve the loss of a limb, the source of bleeding is pretty obvious — you don't need a degree in medicine to identify that. But there are other kinds of bleeding, which are not quite so easy to localise. This is the case, for instance, when the bleeding is in the gastrointestinal tract, which is the medical term for the stomach and gut. Internal bleeding of this kind can be extremely dangerous, but this isn't always the case.

We drive to an estate of pre-fab tower blocks and quickly find the address. We park, and I jump out and slide open the side door of our ambulance, where our emergency rucksacks are waiting to be shouldered. Fully loaded, we tramp up the stairs. I pull on a pair of latex gloves as a precaution, and

70

Thomas does the same. Arriving slightly out of breath at the top of the stairs, we find a woman waiting at the front door of her flat.

'Hello, I'm Johannes von Borstel, and this is my colleague ...'

The woman interrupts — 'My husband is in the bathroom. He's throwing up blood!' Understandably, the woman seems upset.

'I'm puking blood in jets!' a deep male voice resounds from inside the flat.

We follow the woman into the bathroom, where her husband is kneeling in front of the bathtub, supporting himself on his arms, his pale face looking downwards. The sides of the bathtub are streaked with smears of blood. We approach the man and begin the standard questioning procedure, known by the acronym SAMPLE. What are his Symptoms? Does he have any Allergies? Has he taken any Medication? Does he have any Past medical history relevant to this episode, or has it ever happened before? When was his Last meal? Were there any particular, relevant Events leading up to this episode? Asking all this helps paramedics gain a good overview of the medical situation within a short space of time.

While we carry out the SAMPLE history assessment, Thomas prepares an infusion and I measure the patient's blood pressure and pulse rate and try to gain a general impression of his condition. The SAMPLE history revealed no additional abnormalities, but the patient does inform us that he had a stomach ulcer five years earlier, now healed. Despite his pallor and his low blood pressure, the man seems quite coherent and alert. Yet from time to time, he's overcome by uncontrollable attacks of retching. The one thing that surprises me is that there's nothing in the bathtub apart from a few streaks of blood.

'Where were you when the vomiting started?' I ask. 'I need to see it.'

'Are you sure you really want to?' asks the man, as the blood-smeared corners of his mouth twist into a wry smile.

'I think I'd better.'

The woman leads us to the kitchen, where a small pool of blood, about 15 centimetres across, awaits us on the floor. Dark-red blood, with even darker, bean-like lumps in it. I check that they are not really beans by touching them. They're lumps of coagulated blood!

I return to the bathroom, where Thomas has now inserted a catheter in the patient's vein, and the infusion is running and exerting its stabilising effect on his circulatory system. The colour returns to the man's face, and he even asks if he can stand up, but Thomas persuades him that it might not be a good idea. The risk of him finding himself horizontal again straightaway is too great.

We have a decision to make. Should we wait for the emergency doctor to arrive, or take the patient straight to hospital? Transporting a patient without a doctor present can be risky, but waiting too long for a doctor to turn up can also be a risk. The doctor is going to take some time to arrive, whereas we can be in the ambulance within three minutes, and at the emergency ward of the nearest hospital in four more. We set off for the hospital.

Within 15 minutes of arriving, we conclude the handover to the duty doctor there. A good, smooth, routine call-out. On the way back to the ambulance station, there's just time to pop in at a bakery. As we're perusing the cakes, our pagers begin to beep again.

'We'll be back later,' says Thomas with a sigh.

Another emergency. Male, 53 years of age, with suspected gastrointestinal bleeding and vomiting. Is this deja vu? Is it the same emergency, reported again by mistake? Not quite: the address and age are different. So it's quickly on with the siren and blue light, and off we go. Luckily, the roads aren't as busy now, and I start to ponder. The patient's name and address seem suddenly familiar.

'Have you heard that name before?' I ask Thomas.

'I was called out to a CDI* with seizures there recently,' comes his prompt reply. Then it comes to me in a flash. I've been called out to that address before, too, but to deal with a head wound connected with domestic violence and alcohol. The police were also there; in fact, it was the police who called us out.

Indeed, when you work as a paramedic, there are some households you have to visit more often than others. This is one of them. A man in his 50s lives at this address with his

---

*   Short for 'combined drug intoxication', i.e. after taking several substances at once.

wife. We'd never know what he or she might confront us with each time they called; like some sort of Kinder Surprise for paramedics. Only with no chocolate and no toy.

As soon as the ambulance comes to a halt, all the moves are more or less automatic. Door open, rucksack and oxygen cylinder on the shoulders, and off to the front door. Here, again, we are met by the patient's wife. She looks extremely upset.

'Quick! My husband's in his chair in the living room, throwing up blood.' When we reach the living room, however, he's no longer in his chair, but lying face-down on the floor. Behind him is his armchair, and next to him an up-turned bucket from which a puddle of blood is seeping onto the carpet. The blood is liquid, with no visible lumps in it.

The patient is unconscious and no longer breathing. We immediately start resuscitation procedures: clear his airways, intubate him, and begin cardiac massage. We do everything we can, including administering adrenaline and atropine. But without success. After a few minutes, the emergency doctor arrives. We continue resuscitation procedures together. Yet no matter how much we do, the man never regains consciousness. Eventually, all the doctor can do is pronounce the patient dead.

It might seem strange that these two cases, which appeared so similar at first glance, should have had such different outcomes. Both patients were vomiting blood, but there was a vital difference in where that blood had come from. This is a fundamental problem: when you're presented with a patient vomiting blood, there's initially no indication of where the blood originated. Of course, it always exits the body via the patient's mouth, but you can never tell where inside the body the leak will be.

The most probable sources of the blood are the stomach, the gut, or the oesophagus, but it may also have run down from the inside of the nose through the throat and into the stomach. Knowing the origin of the blood in a patient's vomit is crucial for assessing the urgency of the case and how precisely to proceed with treatment, so it's important to find out this information as quickly as possible.

The best way is to examine the consistency of the vomited blood. In our first case, it contained lumps; in the second case, it didn't. This is rather an important difference, because blood contains proteins that can agglutinate and form clumps under certain circumstances, such as coming into contact with gastric acid over a longer period. This can happen if the patient is suffering from a stomach ulcer, most commonly caused when the lining of the stomach is colonised by *Helicobacter pylori* bacteria. These tiny microbes cause an inflammation of the stomach lining, which is then no longer able to protect the stomach's wall from the effects of its own acid secretions. This results in lesions that bleed slowly into the stomach.

Experience has shown that some patients struggle repeatedly with stomach ulcers. The risk of their condition recurring — medical professionals call it relapse or recidivism — is high. This can be due to a genetic predisposition, but it can also be caused by smoking, drinking alcohol, or taking certain medications. For anyone who regularly takes aspirin over a prolonged period of time, for example, the risk of developing an ulcer is four times higher than normal.

When the blood from the ulcer comes into contact with gastric acid in the stomach, it coagulates, and the vomited blood takes on a very characteristic consistency — reminiscent of coffee grounds, as in our first emergency. Indeed, after tests in hospital, that patient was diagnosed with a recurring stomach ulcer.

The daily life of the second patient was one of drug and alcohol abuse. This is a lifestyle that causes damage to the entire body, but to the liver especially. Regular alcohol consumption can seriously damage the liver, and, in the most severe cases, this results in cirrhosis. In this condition, more and more of the cells of the liver die off and are replaced by connective tissue. This gives the organ a compact, knotty structure. Although it can also be caused by a viral infection, in developed countries around half the cases of liver cirrhosis are due to alcohol.

Cirrhotic thickening of the liver tissue increasingly prevents blood from flowing through the organ, causing it to back up in the portal vein, a blood vessel that gathers nutrient-enriched blood from the gut and carries it to the liver. This goes on until connections — once again called anastomoses — grow between the portal vein and the superior vena cava, via which blood flows directly to the heart without passing through the liver.

Such anastomoses continue to grow in several parts of the body, including around the oesophagus, where the increased blood pressure (due to the back-up of blood in the portal vein) causes them to swell and eventually form thick varicose veins (oesophageal varices). These can then fatally burst, pumping large quantities of blood into the oesophagus, which then flows into the stomach. From there, it's soon ejected as projectile vomit. This explains why what looks like gastric bleeding can actually be caused by alcoholic liver disease.

In this way, reading coffee grounds on the ground, or, more accurately, examining the structure of bloody vomit and the lifestyles of the two patients, provided vital clues as to the origin of the blood. Whether the first patient's ulcer was partly caused by drinking is difficult to say. However, there's little doubt that the fatal bleeding suffered by the second patient was a direct result of many years of alcohol and other drug abuse.

Of course, we enjoy alcohol, tobacco, and all sorts of other addictive substances. And enjoyment is good for the heart! However, we should learn to recognise when we are doing damage to ourselves. We could simply do without that after-work beer, stay off the hard stuff when we're out drinking, and quit smoking. Anyway, after all the Marlboro Men died of heart disease, lung cancer, and other smoking-related conditions, smoking at least has become pretty uncool.

Gridlock in the Heart

Coronary heart disease, arteriosclerosis, and heart failure

Total Closure

Whether we like it or not, we age as we progress through life. It's inevitable, and there's no way to stop it. The most obvious sign that our bodies are ageing is in our skin. It loses elasticity as we age, we get wrinkles, and that baby-smooth complexion becomes a thing of the past. Less obvious to our eyes is the fact that the same thing is also happening *beneath* our skin. As if the outward signs of ageing weren't cruel enough! Our blood autobahn also gradually loses its elasticity and becomes increasingly porous. As with an actual autobahn, the cause of this is constant daily use. Only, in our blood vessels, it is not 40-tonne trucks that slowly but surely destroy or clog the transport system, but rather unhealthy eating, smoking, heavy drinking, and lack of physical exercise.

We can stretch the autobahn analogy even further. When the motorway is fully closed, a traffic jam will result, and the same happens in clogged blood vessels, although the jam usually builds up over a period of many years or even decades. The worst and often irrevocable result is either a classic heart attack or sudden cardiac death.

When fat and plaque deposits build up on the walls of the coronary arteries without completely blocking them and causing a heart attack, the supply lines to our heart muscles become ever harder and narrower. This continues until the vessels are no longer able to deliver enough oxygen-rich blood to the muscles of the heart when it is under stress. This is called coronary heart disease, and it can cause symptoms of varying severity.

One common symptom is the chest pain known as angina pectoris, which has already been mentioned. It usually comes in sudden attacks, which sufferers describe feeling as if someone were pulling a belt extremely tight around their chest. They can barely breathe and often begin to panic. Which isn't surprising, since everything seems to point towards an imminent heart attack. But the pain and other symptoms soon subside and everything returns to normal. However, this doesn't mean everything is alright. An angina attack is in fact a clarion call, signalling that the coronary arteries are already severely negatively affected.

Unfortunately, it's virtually impossible to treat the causes of arterial narrowing. And the condition doesn't get better by itself. On the contrary, our 'lifestyle 40-tonne trucks', such as smoking and unhealthy eating, ensure that our blood vessels become progressively more clogged and impassable.

Once this process has begun, then, if not halted by a massive lifestyle change, it is only a matter of time before acute coronary syndrome develops. Acute coronary syndrome refers to a group of cardiovascular conditions and their symptoms caused by narrowed or blocked blood vessels. They include unstable angina and heart attacks, but cardiac arrhythmia (irregular heart beat), heart failure, and sudden cardiac death may also be related to acute coronary syndrome.

There is no single, clear cause of coronary heart disease, but there are several risk factors that can be positively influenced by treatment. These include diabetes, high blood pressure, and increased blood-fat levels. Smoking and, not least of all, lack of exercise are of course also deadly factors.

Hardening of the arteries can lead not only to heart attacks, but also to strokes and even vascular dementia,* which increases with age. Alongside the many advantages of old age — boundless wisdom, the time to spend the entire day gawping out of the window, hanging around on cruise ships rather than in offices, and the right to take absolutely ages to answer the telephone — the heart harbours a couple of not-so-nice surprises for our autumn years.

---

* A stepwise cognitive decline caused by problems in the supply of blood to the brain.

A Tight Squeeze

Ladies and gentlemen! Narrowing or hardening of the arteries has a name. Let me introduce you to the bane of the human race — arteriosclerosis. We are now going to examine this problem closely, since it is not a malicious bacterium nor a malevolent virus, nor a biological weapon, nor even the Kardashians that are the scourge of most people on the planet, but rather the simple narrowing of their blood vessels. There is barely another disease that is so widespread.

The cruel thing about it is that it creeps up on us over a period of decades without our noticing the slightest effect until the situation is already critical and symptoms finally begin to make themselves felt. In fact, our blood vessels begin to clog due to deposits of plaque and fat on their inner walls when we are only about 25 years old. And the process continues throughout our life until problems such as angina eventually make us aware of the encroaching evil.

Some time ago, the Quebec Heart and Lung Institute and Laval University published a joint study of 168 male and female subjects between the ages of 18 and 35 who, without exception, had none of the risk factors for cardiovascular disease. Using magnetic-resonance imaging, the scientists looked for fatty deposits in the subjects' chest and abdominal areas, and examined the state of their carotid arteries, since that's the best place to identify early signs of arteriosclerosis.

And lo and behold: they found that even young, apparently vigorous subjects were already affected by arteriosclerosis. So if anyone was thinking there is no need to worry about the condition in the first third of their life, I'm afraid I must disappoint them.

How is it that the human body, which has managed to adapt really rather well to constantly changing conditions over hundreds of thousands of years of evolution, has no defence against this disease? The answer is simple: arteriosclerosis has only become relevant in the last hundred years. One of its causes, and one that should not be underestimated, is the increase in life expectancy in recent years due to advances in medical science. Our bodies now have much more time than they did in the past to develop arteriosclerosis.

In the Middle Ages, the average life expectancy stood at around 30 years. Of course, people suffered far less from cardiovascular problems during their short lifetimes — on the other hand, there were more epidemics and what we now think of as 'childhood diseases' to send people to an early grave.

In addition, our lifestyle and dietary choices are far more conducive to the development of arteriosclerosis than those of people 600 years ago. Our modern diet is considerably richer in sugar and fat (and excess sugar is transformed metabolically to fat within the body). This results in much of that excess fat being deposited on the walls of our blood vessels.

There are several theories as to how arteriosclerosis manages to spread through the body. In 1976, the American pathologist Russell Ross formulated the response-to-injury hypothesis, which is best illustrated by a comparison with a medieval fortress under attack.

Imagine your body is a fortress with many inner chambers filled (by unhealthy eating) with fat, and the inner walls of your blood vessels are the protective walls surrounding the fortress. Enemy knights such as Sir Bacterium and Sir Virus are continually trying to capture your good body fortress. They're not interested in the fat inside, but they are constantly on the rampage all over the fortress, especially the castle walls, gradually destroying the fortress altogether. This is exactly how the toxins from bacteria or viruses attack and injure the inner walls of our blood vessels. And, according to the response-to-injury hypothesis, every case of arteriosclerosis begins with just such an injury.

And how do you destroy the walls of a besieged castle? With a battering ram! If the attackers have a battering ram in their arsenal, the castle walls will give way all the sooner. The same is true of the walls of our blood vessels. Only, in their case, it's not a battering ram that does the damage, but mechanical stress and high blood pressure. If part of the wall is about to cave in or has already been destroyed, the inhabitants of the castle must of course react. Messages of the impending doom immediately spread throughout the entire fortress.

The messengers within our bodies are growth factors

and cytokines.* They ensure that particularly steadfast inhabitants of the castle, the cells of the blood-vessel walls, proliferate in the middle layer (tunica media) and migrate to the inner layer (tunica intima). Hot on their heels come the macrophages, or scavenger cells; usually very useful castle inhabitants. When attracted by damage to the inner vessel walls, they greedily start tucking in to fatty deposits and gobbling them up. Then the cells of the vessel walls also begin to take up this fat.

As with fans of fast food, the constant eating causes these fat-laden macrophages and vessel-wall cells to change their appearance. And when someone changes their appearance, they often get stuck with a nickname. Noticing I had gained a few pounds since our previous meeting, a friend of mine recently nicknamed me Fatty MacMunchbin. Mean, but quite funny I have to admit. Scientists were not quite so mean when naming the fat-laden cells. Rather than calling them Fatty-MacMunchbin cells, they refer to them as 'foam cells'. Looking at them under the microscope, the reason becomes obvious. Following their feeding frenzy, their insides look as if they are filled with a rough foam.

Before this point is reached, arteriosclerosis can be reversed. This was found to be the case in individuals who began an endurance training regime, for example. Such subjects were found to show a significant reduction in their blood-cholesterol levels and, most importantly, an improvement in the ratio between 'good' and 'bad' cholesterol

---

* Proteins that influence the development and growth of cells.

in their blood. Once foam cells have formed, however, this fatal process appears to be almost completely irreversible.

This is probably due to the fact that the proliferation and migration of the cells of the blood-vessel walls, as well as the formation of foam cells, over an extended period of time cause those tissue changes so typical of arteriosclerosis: plaques. But, according to the 'response-to-injury' hypothesis, the primary cause of this is always an injury to the inner wall of the vessel.

A different theory of how arteriosclerosis develops was put forward by the American researcher and Nobel Prize–winner Joseph Leonard Goldstein in 1983. He was the first to report that macrophages take up a chemically altered protein known as oxidised LDL,* and only then do they turn into foam cells. According to his 'lipoprotein-induced arteriosclerosis' hypothesis, the process begins with the modification of LDL — the injury to the blood-vessel wall comes later. But Goldstein and Ross agree on one point: both believe that foam cells are to blame for triggering a massive inflammatory response.

Inflammation is in fact a clever defence strategy on the part of our bodies, protecting it from invaders such as pathogens. If you graze your knee, there's a very good reason why it swells up, hurts, turns red, and gets warm — the so-called 'cardinal signs of acute inflammation'. They occur because our body immediately increases the blood flow

---

* LDL = low density lipoprotein. For more on this, see 'Should the Easter Bunny Go Vegan?', p. 118.

to the area of the injury, so that as many immune cells as possible arrive at the site as quickly as possible to defend the body against pathogens and to reseal the wound. This increased supply of blood is what causes the wound to get warm and go red.

The pain we feel — also an extremely useful reaction — prompts us to take particular care of the inflamed area and to keep it stable by not moving it too much. So in the case of a grazed knee, inflammation is a good thing. But when blood vessels get inflamed, the fatal result is the creation of deadly plaques.

If the inflammation spreads deeper into the blood-vessel wall, it promotes a gradual alteration of the tissues there. Connective-tissue structures called fibrous caps begin to form in the vessel wall, and this in turn promotes the development of blood clots, also called thrombi (singular: thrombus). In a least-worst-case scenario, these thrombi close the blood vessel at the location where they develop; in a most-worst-case scenario, they're swept along with the bloodstream and eventually block a blood vessel somewhere else in the body.

If this happens to be a coronary artery, the result is a heart attack. If it happens in one of the blood vessels of the brain, the result is a stroke. And if the thrombus blocks a blood vessel in the lungs, doctors speak of a pulmonary embolism. These are the most-serious consequences with which arteriosclerosis suddenly makes itself known.

The fibrous alteration of the artery's tissue makes it

porous and prone to the build-up of calcium particles, and the vessel wall becomes not only thicker, but also harder. This explains the common term — hardening of the arteries — which, however, describes only a small part of what happens inside the blood vessels. But 'hardening of the arteries' is easier to remember than 'LDL-foam-cell-forming-fibrous-cap-emplacement-vessel-degradation'.* But whatever we call this terrible condition, it's important to know that it is responsible for a whole range of problems.

---

* Even this made-up term falls far short of describing the entire phenomenon.

A Big Heart

Cause of death: heart failure. Those words are found on many a death certificate. But what do they actually mean? In fact, heart failure is a rather meaningless term used by doctors when they don't know the precise cause of the cardiac arrest that led to death. In such cases, medics speak of (acute) decompensated heart failure, and it's one of the most common reasons for hospitalisation. It is often associated with other medical conditions and is usually the result of coronary heart disease. It is particularly common among people with type 2 diabetes.

Heart failure occurs when the heart is no longer strong and efficient enough to properly do its job of pumping an adequate supply of blood and oxygen to all parts of the body. In most cases, it's due to arteriosclerosis causing the coronary arteries to narrow, making them unable to supply the muscles of the heart with enough blood. High blood pressure is another cause of this insufficiency, since if blood encounters increased resistance as it flows through the arteries, the heart is forced to work harder than normal; in fact, it faces a real slog.

Such an increased workload is not without long-term consequences: our central blood-supply organ gets increasingly weaker — just as we do when we work too much and end up with executive burn-out. Heart failure is particularly common in people aged 70–80, and men are affected more often and earlier in life than women.

Furthermore, an abnormally fast or slow heart rate, an irregular heartbeat, heart-valve defects, or a so-called cardiac or pericardial tamponade (when fluid, for example blood, builds up in the pericardium — the sac that surrounds the heart — compressing the heart and reducing its ability to function properly) can all lead to heart failure over time. It can also be due to an inflammation of the heart muscle (myocarditis), a pulmonary embolism, or a non-fatal heart attack. Phew! That's a long list of diseases and conditions that can compromise the heart's pumping ability. And that list is not even complete. There are other possible reasons why a heart may lose its normal robustness.

One of them is anaemia, a decrease in the number of red blood cells floating around in the blood, which means it's able to transport less oxygen. This in turn means the heart has to work harder to compensate. To help it, the body releases adrenaline and noradrenaline, hormones that cause the heart to beat faster. In addition, the renin-angiotensin-aldosterone system — an enzyme and hormone system that the body can use to regulate blood pressure* — increases the overall volume of blood, thus increasing the pressure within the blood vessels. These effects serve principally to make sure that our vital organs continue to be supplied with enough blood. Yet sooner or later, they begin to harm the heart more than they help it. Eventually, the heart itself grows in size in

---

* See also, 'Bottles on the Lawn', p. 237.

order to increase its pumping power.

Like any other muscle, the heart increases in size when it's exposed to a higher workload. When this happens to the heart of an endurance athlete, whose muscles need large amounts of oxygen, it is a normal development and not in itself dangerous. But it's a different story if a person's heart becomes enlarged due to disease factors. If the pressure inside the heart is too high, the organ becomes enlarged, and continues to increase in size. The heart turns into The Hulk!

This increase in bulk means the heart itself requires an ever-increasing supply of oxygen, which in turn causes it to swell even more: a deadly vicious circle. In addition, as it grows excessively in this way, the heart lays down connective tissue along with muscle — a process called fibrosis. As with other fibrotic changes to an organ of the body, such as cirrhosis of the liver as described earlier, the effect of this is to slowly but surely reduce the ability of that organ to function properly.

Finally, an overactive thyroid over a long period of time can also cause heart failure. This is because the hormones produced by this little gland in the neck cause the heart to beat faster. (Above 100 beats per minute, doctors call this tachycardia.)

Heart failure is basically made up of two components: the systolic dysfunction (systole: contraction of a ventricle) and the diastolic dysfunction (diastole: relaxation of the ventricle). Systolic dysfunction is a failure of the heart's pump action, or, more precisely, that of the left ventricle. Diastolic dysfunction is the failure of the ventricle to adequately relax and fill with enough blood. In both cases, the result is that the weakened heart pumps too little blood into the circulatory system, and the entire body is insufficiently supplied with nutrients and oxygen.

Just as anatomists divide the heart into a right and left side with different functions, doctors also classify types of heart failure according to the side of the heart that's affected. If it affects mainly the muscles of the right ventricle and atrium, whose job it is to pump carbon-dioxide-rich blood to the lungs, then medics speak of right-sided failure. If the right side of the heart is no longer able to pump the blood that flows into it, there will be a tailback in the veins of the body. In response, this side of the heart tries to work harder to get more blood to the lungs, resulting in a thickening of the walls of the right chambers of the heart. Eventually, the heart no longer has the strength to maintain these valiant efforts. One sign of this is swollen veins in the neck, which can be felt or even seen. (Again, just like The Hulk. Only we don't go green.) Another sign of right-sided failure is an accumulation of fluid in the legs and abdomen.

Right-sided heart failure is a right convivial condition, but a bad guest. It seldom comes alone, preferring to bring a couple of uninvited friends along. Unfortunately, these gate-crashers are badly behaved, too. Right-side heart failure usually turns up accompanied by, or as a result of, a reduction in the pump action of the heart's left side. Under normal circumstances, this is the side that receives oxygen-enriched blood from the lungs and pumps it round the rest of the body. If its pump action is no longer strong enough, there will be another tailback, but this time to the lungs. The medical term for this is pulmonary congestion. The dangerous thing about such pulmonary congestion is that, as the pressure in the vessels of the lungs increases, fluid is pressed into our respiratory organ. The result is that the lungs are slowly but surely flooded with liquid.*

Failure of both sides of the heart is called biventricular failure. Often developing as a result of another condition or illness, biventricular failure can occur within days or hours — 'acute failure' in medical parlance. Gradual failure of the heart over months or even years, on the other hand, is described as 'chronic heart failure'.

The New York Heart Association has published a table classifying heart failure into four degrees of severity. According to this table, Class I heart failure involves no limitation of the patient's physical activity; there are no symptoms when the patient is at rest or under normal daily physical stress. Slight limitation of physical activity comes with Class II heart failure. If the patient is experiencing a marked limitation of physical activity, and even a less-than-ordinary level of activity causes palpitations, breathing difficulties, or tightening in the chest, then Class III has been reached. Class IV describes a condition in which the patient is unable to carry on any physical activity without

---

* Pulmonary oedema, as described in 'Of Dragon Boats and (Oar) Strokes', p. 49.

discomfort, and requires help with day-to-day tasks.

Assessing the severity level of a patient's heart failure helps medical staff to plan the best course of treatment. As you might expect, medication such as antihypertensives (drugs that lower the blood pressure) and diuretics (to remove fluid from the body) can increase the patient's quality of life. But prescribing pills only makes sense if patients are also willing to adapt their lifestyle to their medical condition. This includes giving up smoking, and drinking less — or even better, no — alcohol.

In addition, reducing dietary salt intake will help make the heart's work easier; salt promotes water retention in the body, causing the overall volume of blood in the body to increase, thus forcing the heart to work harder. This doesn't mean, however, that heart-failure patients should reduce their fluid intake. Unless instructed otherwise by their doctor, they should drink at least two litres of water a day. If they also eat a wholesome diet that promotes cardiac health, and shed a bit of excess weight, they can live with their heart condition and enjoy a good quality of life.

I can well imagine that shifting one's dietary habits and losing weight is difficult at first. They are two things that don't sound like fun at all. But once a start has been made, the rest often follows automatically. One possible way to shift one's diet is literally to shift one's food. I mean that seriously.

A friend recently told me he was shifting his diet. Instead of keeping his chocolate supply to the left of his computer keyboard, he now keeps it on the right. Not a bad idea, I suppose, but not particularly effective either. However, what if he shifted his stash of sweets to the cellar or the attic? Keeping them at a distance would at least mean less temptation to snack out of boredom. That would certainly be a good start. The same friend recently sent me a text saying he had just burned 800 calories. Attached to the message was a photo of a charred pizza. What a joker!

Eating to Your Heart's Content

The connection between diet and a healthy heart

Omega-3 or Not Omega-3, Fat Is the Question!

There lies the object of my desires in all her glory, tempting me. The air is heavy with my temptress's scent, so enticing, I can hardly control myself. With deliciously gleaming skin, my lover's curvaceous shape takes my breath away. This is a real heartbreaker, lolling before me, naked and hot. And cheap. Just four euros, to be precise.

Then the voice of my conscience chimes in. 'Johannes! No! She's no good for you. Don't do it!' Great, thanks a lot! My conscience is a real killjoy. Every time I pass a sausage stand, the same drama plays out in my mind. I'm a great lover of sausage with curry sauce and chips. Unfortunately, like so much that tastes or feels good, this delicious treat is pretty bad for your health.

We're confronted with an abundance of food every day, meaning there's a great risk that we will stuff ourselves with unhealthy fare. But how can we keep track of all this food and its implications for our health? How do I know what I should be eating and when I can relax those strict healthy-eating rules and treat myself a little? Although medical care is improving all the time, cardiovascular disease is still on the rise in the developed world.

This is a real 'first-world problem'. Many times,

cardiovascular disease results from eating wrongly or simply too much. What we eat often makes us feel full, but it also makes us sick in the long term. Preservation methods can rob our food of many important nutrients; food chemistry can leave it lacking in essential vitamins and minerals. All that would not be so bad if it weren't for the fact our cardiovascular system suffers so much from such a poor diet.

Processed and convenience foods are the biggest problem. Unfortunately, they're also the biggest favourites. It's understandable, since who these days has the time or the desire to cook for themselves? After a busy day at work, or during a short lunchbreak, a steak sausage sandwich from the snack bar next-door is much more appealing, and more importantly, quicker and easier, than slaving away over a hot stove. But, in fact, eating well for a healthy heart is not as time-consuming as people may think, and it can be enjoyable.

The first step in the right direction is: choose good fats! Many people believe fat is unhealthy per se. But this isn't completely true, since fats can be divided into different categories, ranging from 'good' fats, such as those found in linseed oil, for example, to 'bad' fats — because they are hydrogenated — such as those found in margarine and coconut and palm oil. Unfortunately, it's the unhealthy fats that are mostly contained in convenience foods. They play a large part in making inflammations in the body worse, putting our cardiovascular system under great pressure. So choose good fats for a healthy heart!

But what exactly makes some fats healthy and others

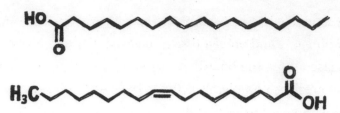

Stearic acid (top) is a saturated fatty acid and does not contain double bonds. Oleic acid (bottom), by contrast, is an unsaturated fatty acid and has one carbon–carbon double bond.

unhealthy? To understand it, we need to realise that some fatty acids are saturated, while others are unsaturated, and the difference is in their chemical structure. Saturated fats mainly hide out in animal products, including butter, cream, and bacon. Is your mouth watering yet? Sorry, but these fats are famous for increasing the levels of cholesterol in the blood. One type of food that is low in saturated fats, on the other hand, is lean meat — in particular, chicken and turkey meat; but also fish and seafood.

In 2007, the German Nutrition Society published guidelines, based on academic studies, on which types of fat have positive or negative effects on our cardiovascular system. According to that information, the risk of coronary heart disease falls by 19 per cent if saturated fats are replaced with polyunsaturated fats,[*] ideally by unsaturated omega-3 and omega-6 fatty acids.[†] Yet the more you read on this topic,

---

[*] For those interested in the chemistry here: polyunsaturated fatty acids contain multiple double bonds in their carbon-hydrogen chains.

[†] The number indicates between which carbon atoms the last double bond in the chain is located.

the more confusing it seems to become. While some studies say omega-3 and omega-6 fatty acids lower the risk of heart disease across the board, others recommend that omega-6 fatty acids be completely avoided by heart-disease patients. Whose advice are we to follow?

What is certain is that several positive effects have been ascribed to omega-3 fatty acids. Many studies agree that they increase the elasticity of the hair and skin, boost the immune system, and help fight harmful inflammation. Most importantly, they protect the heart, since they have a positive effect not only on the cholesterol contents of the blood, but also on blood pressure and sugar levels.

Omega-6 fatty acids are thought to suppress the positive effect of their omega-3 colleagues, however. But it should be said that the various studies are so contradictory that they need to be approached with caution. As is so often the case in such debates, the truth probably lies somewhere in between the two extremes. It probably comes down to finding the right balance.

There seems to be little doubt, however, that the anti-inflammatory chemical-transmitter substances made from omega-6 fatty acids are less effective than those based on the omega-3 fatty acids found in oily fish like tuna, mackerel, salmon, and herring. Seafood, such as mussels, also contains large amounts of omega-3. For this reason, nutritionists recommend that the ratio between people's dietary omega-6 and omega-3 intake should not exceed 4:1 (i.e. we should not eat more than four times as much omega-6 as omega-3).

Unfortunately, this tends to clash with our Western eating habits. We all consume around 10–20 times as much omega-6 as omega-3. This is due to the fact that omega-6 fatty acids are contained in all our favourite foods, like animal fats, meat, dairy products, and salad dressings.

This advice may be well-meaning, but it's not so easy to follow. What was it again? Four times more omega-6? More than what? Which kind is it that's in chips again? Dietary advice is often rather vague and difficult to remember. And more to the point, almost impossible to live by. Clear, easy-to-follow tips are few and far between, not least of all because everyone has their own individual metabolism, body proportions, and pre-existing medical conditions. For this reason, dietary advice should mostly be seen as a set of rough guidelines. Furthermore, most of us do not have time to spend examining our dietary habits in such detail. Nutrition is a complex science.

For those who really want to do the right thing, there's the option of consulting a nutritionist or dietitian who can draw up a personalised dietary plan for you. But even those who don't have that option can do their hearts a favour whenever they eat and drink. None of us can change our age, gender, or genetic predispositions, but all of us can alter our eating and drinking habits to counter high blood cholesterol, diabetes, obesity, and high blood pressure[*] — for example, by replacing

---

[*] A combination of obesity, high blood pressure, high cholesterol, and high blood-sugar levels is known as metabolic syndrome or 'the deadly quartet' (see p. 133). This is a precursor to many vascular diseases.

saturated fats with unsaturated fats as far as possible.

Cutting out fats altogether makes little sense. It is much better to pay attention to the quality and composition of the fats we eat. According to the guidelines of the German Nutrition Society, 80 grams of fat per day is a healthy amount for most adults; and as far as possible, that fat should come from vegetable sources, like rapeseed or soya oil and the spreads made from them, because they contain more omega-3 fatty acids.

But eating the right kind of fat isn't the only thing we need to pay attention to. The role of sugar in our diet is a bittersweet one. Sugar is the main energy supplier for our bodies, but there's so much of it hidden in the food we eat that our bodies have difficulty using up this permanent oversupply of energy. They react to the situation by storing this energy in fat reserves, and that can only lead to obesity. Still, 'energy reserves' does sound more impressive than 'beer belly', doesn't it?

Eat Yourself Fit

'More sugar!' orders my niece. We are in the kitchen making iced tea.

'Er, no … That's enough now. Any more sugar and it will be sickly,' I reply with a long-suffering undertone.

She frowns. 'Huh? What's that supposed to mean?'

'Well, it'll be so sweet it will make you sick. If you want more sugar, have a bite of this.' With a grin, I offer her a potato.

'Oh maaan,' she sighs. 'I'm not stupid! Potatoes aren't made of sugar. That would taste disgusting!'

My niece is right about one thing: biting into a raw potato *is* disgusting. But I would have to differ with her over whether there's any sugar in a potato, since the sweet stuff is a cunning master of disguise. 'Sugar' is a generalised term for a large number of more- or less-sweet-tasting compounds called saccharides. They are roughly divided into groups — including monosaccharides, such as glucose and fructose; and disaccharides, such as lactose, maltose, or familiar household sugar, sucrose. As you may guess from the name, monosaccharides are simple or single sugars, consisting of one fundamental unit of sugar, while disaccharides (double sugars) are made up of two such units. Sugars containing up to ten units are called oligosaccharides, and those which are more complex still are known as polysaccharides.

Glucose (top) is a monosaccharide, consisting of just one sugar unit.
Lactose (bottom) is a disaccharide.

The polysaccharides include starch, which we are familiar with from potatoes. Although it tastes only vaguely sweet at best and is almost insoluble in water, starch is nonetheless a sugar compound — more precisely, a long chain of glucose units, which can be broken apart to release energy. Starch is the form in which most plants store sugar.

We humans, on the other hand, store glucose in a different form: glycogen. In this form, the individual glucose units form branching rows, each one holding onto the next, almost like protesters at a demonstration joining hands for unity.

When faced with an overabundance of glucose, the body transforms it into glycogen, mainly in our liver and muscles, and stores it away for leaner times. During periods of extreme exertion, such as a long-distance run, that glycogen can deliver energy after the rest of the body's simple-sugar supply has been used up.

All these different sugars are described by one general term that's familiar to everyone: carbohydrates. They are essential for our survival; our bodies simply could not function without them. The most valuable to our organism aren't the monosaccharides and disaccharides, however, but the long-chained carbohydrates, such as starch. This is because short-chain carbohydrates are rapidly broken down in the gut and absorbed into the bloodstream. With the help of insulin, they enter our muscle cells, where they provide energy. This is why our blood-sugar concentration rises very rapidly when we eat such sugars ... but it also falls again equally rapidly.

Long-chain carbohydrates, by contrast, are broken down into their individual building blocks and absorbed into the bloodstream much more slowly. That's why they provide the body with a much more sustained and steady supply of energy for all the nonsense we demand of it. One source of such long-chain carbohydrates is wholemeal bread. At the gym, I often see people chugging sugary juices or even soft drinks. They are undoubtedly a quick and effective energy source during a workout, but only for a short time. A can of Coke provides little power in the long term. This isn't the case with wholemeal bread. It is also made of sugar, but its carbohydrates are arranged in long chains. So, for exertion over a longer period of time, wholemeal bread is by far the better source of energy.

The nutritional information on food packaging includes data on the amount of carbohydrate and sugar in the product. For example, 100 grams of the pre-baked wholemeal-bread rolls I'm eating for breakfast as I write this contain 42 grams of 'carbohydrate', of which 3.2 grams are 'sugars'. The amount of 'carbohydrate' refers to all the sugar compounds in my breakfast rolls, while the 'sugar' part covers only monosaccharides and disaccharides such as refined sugar, fructose, and lactose. Those who want to be kind to their heart and body should take care to eat as little of those types of sugar as possible.

That's easier said than done, however, because sugar is like a drug to our bodies. It triggers the release of the 'happiness hormone' dopamine, which, as we saw earlier, plays a vital part in our brain's reward system. Just like a junkie looking for a fix, our body demands increasing amounts of sugar at ever-shorter intervals — preferably in the form of chocolate or other sweet treats — if we don't keep it in check from the outset. We are the dealer and we have the power to decide: will our best customer get an apple or a cream cake today?

Despite the fact that the term 'sugar addiction' is only now entering general parlance, experiments on rodents, for instance, have shown the animals display abnormal, addictive behaviour when they are denied their sugar fix. In one experiment, scientists fed rats a sugar solution at regular times every day. They then put the long-tailed critters on a normal diet with no sweet treats. The test animals developed real withdrawal symptoms such as lethargy, restlessness, and anxiety. Some even had such strong cravings that they got the shakes. Cold turkey in the rat cage — not much fun.

Although I haven't quite reached the stage where a lack of sweets triggers a response like theirs, I can certainly understand those poor lab rats. Anyway, this means we can totally cut refined sugar, corn syrup, and soft drinks out of our diet. Concerning honey: as a natural product, it's often ascribed healing effects and is recommended as a treatment for coughs and colds, but there's no scientific proof of such a 'healing effect'. Best to keep away from it, too!

Another way to reduce sugar intake is to cut down on food made with white flour.

Although we tend to think of arteriosclerosis as a modern phenomenon, studies have shown deposits in the blood vessels of the millennia-old mummies of Egyptian high priests and rulers, and even in the remains of a princess. Various explanations have been put forward for the cause of these deposits. Smoking can be excluded, as it wasn't a habit among the Ancient Egyptians. Their diet was generally low in fat, and people were much more physically active than we are today. One possible cause of the deposits in the mummies' blood vessels could be eating lots of meat, but it could also be the white-flour products known to have been popular among upper-class Egyptians.

Unlike wholemeal flour, white flour contains next to no dietary fibre, and is made up almost exclusively of carbohydrates — so, basically, sugar. And we now know that this not only increases the risk of developing diabetes, but also is one of the main causes of deposits on the blood-vessel walls, and therefore also of cardiovascular disease.

The heart isn't the only part of the body that's negatively affected by white flour. Researchers have found that people who prefer white-flour foodstuffs and eat them often are more likely to suffer from eye diseases like macular degeneration. In this condition, the cells of the retina inside the eye gradually stop working. In addition, people who banish sugar and white flour from their diet suffer significantly less frequently from gallstones. I could go on listing the woes that are associated with white flour, and the only possible conclusion is: replace white flour with wholemeal as much as possible!

There are some things that might ruin your appetite for flour altogether. Have you ever heard of 'bad lectins', for example? Lectins are proteins contained in flour whose effects include thickening the blood. Of course, this increases the risk of heart attacks and strokes. However, it must be said that lectins owe their bad reputation to a study that was carried out with quantities of the protein that would be extremely difficult for anyone with a balanced diet to take in naturally. Which brings us to the core issue of healthy eating.

The most important thing about diet is — balance.

So cutting out some flour products isn't a bad approach, and replacing white flour with freshly milled wholemeal flour as often as possible is a good idea. The flour should be freshly milled not only because wholemeal flour spoils quickly, but also because the nutrients it contains begin to react with the oxygen in the air immediately after it's milled, destroying many of them.

Ideally, we should get most of our energy from fruit and vegetables rather than flour products anyway. This is a win-win situation, because fruit and veg also have a large number of positive effects on our heart, cardiovascular system, and, indeed, our entire bodies.

There's one fruit — a berry, in fact — that is particularly close to my heart: blackcurrant. Firstly, because we have almost the same name: the German for blackcurrant is *Johannisbeere*. And secondly, because I've made it my mission to have a similar effect on people's health. Blackcurrants have been proven to protect the heart and blood vessels. Just like bilberries (and their American cousins, blueberries), they contain blue pigments called anthocyanins, which have long been used in natural medicine to treat eye conditions, but which also have a positive effect on the cardiovascular system. They act as natural antioxidants, protecting our blood vessels from aggressive free radicals. One such anthocyanin is the pigment myrtillin, which increases the elasticity of the blood vessels.

Watermelon and honeydew melon should also feature regularly on the menu, as the substances they contain have been proven to reduce the risk of blood clots. Watermelon is also said to reduce high blood pressure. It also tastes and looks refreshing and delicious, don't you think?

There's a mushroom that has the same positive effect on the platelets of the blood,* usually known as wood ear mushrooms. You can buy them in good supermarkets and Asian-food shops, often in dried form. Once they've been soaked in water, they can be used as an ingredient in spring rolls, soups, or stir fries, as they take on the flavours of the other ingredients very well. Incidentally, their name stems from the fact that they look like ears when growing on trees in the forest. Notwithstanding possible qualms about eating someone's ears, this mushroom is an inherent and tasty part of Asian cuisine.

Vegetarian food often contains so-called phytochemicals. For instance, pomegranates contain blood-pressure-lowering polyphenols, garlic has sulphides which help prevent thromboses, pulses contain anti-inflammatory saponins, and almost all plants contain phytosterols, which reduce blood-cholesterol levels. If this sounds far too complicated, all you have to do is remember: tonight, it's chickpea and pomegranate salad for dinner. Or bean stew with garlic.

Contrary to widely held beliefs, we don't have to make any special effort to provide our bodies with enough vitamins

---

* Blood platelets, also called thrombocytes, are the blood cells involved in blood clotting. One of their jobs is to help reseal wounds.

and nutrients. All we have to do is what we enjoy doing anyway: mouth open, food in, a good chew, swallow — job done! Three portions of fruit or veg a day is a good guideline to follow, ideally including a range of fruit and vegetables of different colours. This is because it's the phytochemicals in plants that determine their colour.

In order to take advantage of the full diversity of plant foods, it's a good idea to try to eat as colourfully as possible. It is always best to choose fresh, locally produced vegetables, because when they are exposed to light and ultraviolet radiation during long periods of transportation and storage, vegetables lose valuable nutrients. And this means only a fraction of those phytochemicals end up on our plates.

However, shopping for fresh produce every day is time consuming. Those without the desire or time to shop so frequently can happily resort to frozen vegetables. Scientists in Hamburg found that frozen vegetables retain more of their vitamins and other beneficial substances over a period of months than fresh vegetables, even if those were kept in the fridge. They compared the amount of vitamin C in green beans stored for a year at a temperature of minus 18 degrees Celsius with that in beans kept in the fridge. While the frozen beans lost 20 per cent of their vitamin C during the year of storage, that of the fridge-stored beans fell by more than 60 per cent in just a few days.

Those looking for a less 'chilled' way to support their cardiovascular system could do worse than turn to fresh carrots, which have a welcome influence on blood-cholesterol levels. Recognised nutrition guides say even just 200 grams a day are enough to have the optimum effect. Walnuts, oats, and barley also have a significant influence on our cholesterol levels.

Both ginger and garlic help to thin the blood, which of course promotes blood flow through the vessels and improves the blood supply to our organs and tissues. A teaspoonful of grated root ginger in a glass of water is a quick and easy-to-make health drink. Similarly, garlic water, made with two to three teaspoons of grated garlic, will not only thin the blood, but will also have a positive effect on your cholesterol levels — if not on your social life! A useful and almost odour-free alternative is garlic pills, which will not give you bad breath.

And then there's the real panacea: the humble onion. Its curative powers have been known since ancient times. Onions aren't only tasty, they also help thin the blood, thereby reducing the risk of blood clots. And they have a beneficial effect on cholesterol metabolism and blood-sugar levels.

If all this conjures up images of raw vegetables and rabbit food, you needn't worry. Raw vegetables can indeed offer long-term protection for our bodies, but cooked ones can, too. In fact, our bodies absorb some nutrients better from cooked vegetables than raw ones. The antioxidant lycopene found in tomatoes can be taken up much more easily by the body if the tomatoes are cooked in a little oil. The same is true of the provitamins in carrots. Ideally, a healthy diet would include a 50-50 proportion of cooked and raw vegetables. Eating for a healthy heart doesn't mean a boring, monotonous diet — it can be varied and diverse.

Essentially, this is less about doing without, and more about replacing foods with other, better ones. Rather than hydrogenated (hardened) fats, use rapeseed and olive oil; and rather than a creamy dressing on your salad, choose a tasty vinaigrette made of oil and vinegar with herbs. Also, despite the fact that their reputation has been rather tarnished by the likes of chips and crisps, potatoes are not completely off the menu. Switching from chips to boiled unpeeled potatoes can considerably reduce your risk of cardiovascular disease. Two hundred grams of unpeeled boiled potatoes contain just under 0.2 grams of fat; the same amount of chips contains 24 grams, which your body can easily be spared. What's more, although fat is a very effective flavour carrier, you don't need much of it to make a meal tasty, especially if you also combine aromatic herbs and spices for flavour. And if they're nice and fresh, all the better.

When I left home after graduating from high school and had to cook for myself for the first time, I fancied myself as a top-class chef and, with woefully misplaced confidence, already envisaged myself cooking up entire banquets. I now know that I was as far from being a culinary genius as anyone can be. After all, what Michelin chef's repertoire of seasonings only stretches as far as salt and pepper? I was always enthusiastic in the kitchen, but my knowledge of herbs and spices was limited, to say the least. Even today, I can hardly get over the vast array on the shelves of my local supermarket — and how unfamiliar I am with most of them. But, it seems I'm not alone with my lack of knowledge; salt and pepper are still by far the most popular seasonings everywhere.

According to a study carried out by the research institute Euromonitor, the average daily diet in Germany includes around eight grams of salt.* That's a lot, considering, as we've already seen, it really makes sense to cut back on salt for the sake of cardiac health, since too much of it raises the risk of heart attacks and strokes considerably. The German Neurologists' Association even goes so far as to warn that doubling daily salt-intake from five to ten grams increases the risk of a stroke by one quarter. However, all this certainly doesn't mean we should cut out salt altogether, as it is crucial for osmoregulation (maintaining the body's fluid balance). Experts recommend three to six grams of salt (about a teaspoonful) as part of a daily diet. Once again, this is easiest to monitor if you cook your own meals rather than relying on processed convenience foods.

Eating for a healthy heart is an almost inexhaustible topic, so wide-ranging and diverse that I can do little more than offer a few pointers here. This is why I recommend that anyone who wants to delve deeper into the topic should consult a professional nutritionist or dietitian. For those with existing medical conditions, a session with an expert of that kind can

---

* In the US, the figure is nine grams; in Britain, it's seven grams; while in Australia, it's ten grams.

work wonders. For example, if your doctor finds an increased concentration of the particular fats called triglycerides in your blood, then you should know that it's almost certainly because you are overweight. To put it plainly, you need to slim down!

Paradoxically, a high-fat diet increases triglyceride levels in the blood less than a carbohydrate-rich one, which is why it can be useful to eat more 'healthy fat', i.e. fat that's high in unsaturated fatty acids. Losing weight in this way to achieve a BMI of about 18.5–25,* will usually reduce triglycerides back to a normal, or at least only slightly increased, level. This, in turn, reduces the risk of cardiovascular disease. Alcohol also raises triglyceride levels, so for those whose levels are already high, the best advice is to stay off the booze completely, if possible.

Unfortunately, most advice for healthier eating includes going on some kind of a diet. But it should never be a crash diet, lasting a few days or weeks, since most are neither very helpful nor very healthy. An effective long-term solution involves a gradual and, more importantly, well-considered change in personal eating and drinking habits.

Contrary to what the term might indicate, too many hearty meals are not particularly good for the heart. They 'stick' to the heart, or, more precisely, to the walls of the

---

* The BMI, or body mass index, is a value defined as the body mass (weight) of an individual divided by the square of their height. A person with a value below 18.5 is described as being underweight; 25 to 30 is overweight, and anything above that is obese.

coronary arteries, with very serious consequences. When I began to replace old-fashioned rich and stodgy meals with lighter, Mediterranean fare, it felt quite good — hardly a battle at all. On the contrary, I enjoyed eating this way, and still do. Once you have identified hidden sources of sugar, salt, and fats, it's not so difficult to replace them with healthier yet still tasty foodstuffs, especially if you have the support of a professional dietitian or nutritionist. I'll drink a melon shake to that!

Should the Easter Bunny Go Vegan?

Cholesterol: we're all familiar with it from advertisements, especially those showing shiny, happy people merrily eating margarine. Or more precisely: eating margarine that is particularly low in cholesterol, the stuff that's supposed to be so bad for our cardiovascular system.

When I asked friends and acquaintances which foods they avoided because of its cholesterol content, not surprisingly, most said butter and eggs. But is it really true that my yearly hard-boiled-egg binge at Easter might have done me permanent damage? Does the Easter Bunny need a lifestyle change? Should it start hiding carrots in the garden instead of eggs?

Cholesterol is thought of as something that people must ban from their diets entirely if they want to eat healthily, yet cholesterol in itself is vital for the body's survival. One of its functions is to help build and maintain the plasma membranes of every single cell in the body. If a cell membrane doesn't contain enough cholesterol, it can become unstable. Cholesterol also combines with certain proteins to transport signalling substances into and out of cells. Digestion is another process in which cholesterol plays an important part. It is a precursor molecule for the production of bile acids in the liver, which are then stored in the gall bladder. After a rich and heavy meal, those acids are secreted into the small intestine to help in digesting fats there.

If insufficient bile acids are present, the gut is unable to absorb fat and it is secreted undigested. This leads to a condition called steatorrhoea (the fancy medical way of saying 'greasy poo'), which is often accompanied by abdominal pain and flatulence. So you can see: life would be pretty bothersome without cholesterol — and pretty unsexy, too, since the body requires cholesterol to produce sex hormones. This 'stuff of the devil' turns out not to be quite as evil as it's often made out to be.

The late-medieval Swiss physician, and father of toxicology, Paracelsus is quoted as saying, 'All things are poison and nothing is without poison; only the dose makes a thing not a poison.' And he couldn't have been more right. This principle also applies to cholesterol. So why should a substance that is useful to the human organism in so many ways be so unpopular and have such a reputation for being dangerous?

The liver is able to produce almost 90 per cent of the cholesterol our body needs; we have to take in the remainder as part of our diet. Without cholesterol, life would be impossible, but this doesn't mean you can't have too much of a good thing. Raised levels of cholesterol in the blood over a protracted period of time have been proven to be a decisive factor in causing cardiovascular disease. The most serious consequences of cardiovascular disease are heart attacks, strokes, and peripheral-artery disease.

But what exactly does cholesterol have to do with all

that? The best place to begin to explain is with its structure. Whenever I get frustrated about how small my flat is, it makes me think of cholesterol. Why? Because I would much prefer to live in a bigger apartment. Living room, bedroom, kitchen, bathroom, a view of Marburg Castle and the Lahn Mountains beyond — that would be my dream home. Unfortunately, it's likely to remain a dream. To afford the exorbitant property prices in Marburg, I would have to sell everything I own, including probably selling my soul to the Devil.

And this brings me back to cholesterol, since its molecular structure looks just like my dream dwelling. It's made up of carbon rings and chains, a few hydrogen atoms, and one oxygen atom. Their configuration is like the floorplan of my perfect flat: three large rings made up of six carbon atoms each are the bedroom, living room, and kitchen; a smaller carbon ring made up of five atoms is like the bathroom; and beyond that, a view of two mountains — the Castle Mountain in Marburg and the Lahn Mountains beyond. This is a great mnemonic for remembering cholesterol's molecular makeup, and it's useful because cholesterol is important not only for our body chemistry, but also for examiners at the end of biochemistry courses. And, of course, for anyone who is interested in the heart and the diseases that affect it.

As mentioned earlier, most of the cholesterol we need is produced in the liver, but the rest has to come from outside the body, as part of our food intake. And once the body has enough of the stuff, it finds it hard to part with it — like a hoarder who collects so much junk he can hardly live with all

Cholesterol looks like my dream apartment

the accumulated rubbish. To be fair to our bodies, what they really prefer to do is recycle. Much of the cholesterol that enters the gut as a constituent of the bile that helps to digest fat is reabsorbed shortly before reaching the gut's ultimate 'exit', and ends up back in our bloodstream. But it cannot manage this without help. Like a child walking to school, it needs someone to hold its hand.

Certain lipoproteins — high-density lipoprotein (HDL) and low-density lipoprotein (LDL) — help cholesterol on its way through our bloodstream. LDL accompanies it on its journey from the liver to other organs, while HDL helps carry it back to liver again. HDL cholesterol is often referred to as 'good' cholesterol, and its LDL cousin is called 'bad' cholesterol. This is because the liver not only produces cholesterol, but also breaks it down again — and because HDL is responsible for taking cholesterol back to the liver to be broken down, it is seen as the better helper, that is, the 'good' guy.

A condition such as familial hypercholesterolaemia means the number of cholesterol receptors in the liver is reduced, with the result that less of it can be broken down there and so it returns to the bloodstream in its original state. This means the overall amount of bad cholesterol in the bloodstream is increased. If this is compounded by risk factors such as smoking, high blood pressure, or diabetes, there's an increased danger that the cholesterol will settle on the already-damaged blood-vessel walls in the form of arteriosclerotic plaque deposits, which can eventually clog it up completely.

The acceptable maximum amount of bad cholesterol in the bloodstream varies from person to person and must be judged on a case-by-case basis by a qualified doctor. She will ascertain a patient's overall level of cardiovascular risk. The higher this is, the lower the maximum acceptable cholesterol level should be. To put it in hard figures: if a patient has no risk factors or just one, the highest acceptable level of LDL is 4.1 millimoles per litre (mmol/L) of blood.* This represents a low risk to health. If a patient has two or more risk factors, then the maximum amount of bad cholesterol should be 3.4 mmol/L (130 mg/dL). Even then, the heart and blood vessels should beware, because the probability of disease is still increased. People who have already suffered a heart attack or who are diabetic have a high risk of cardiovascular disease.

---

* Alternatively expressed as 160 milligrams per decilitre (mg/dL). This unit of measurement is used in only a few countries, including the USA, France, and Japan, but also parts of Berlin and western Germany.

Their bad-cholesterol levels should not exceed the very low figure of 2.5 mmol/L (100 mg/dL).

Chain smoking, blood pressure as high as a pressure cooker, a history of heart problems in the family, a pre-existing vascular condition, or even a previous heart attack will all put a person in the highest risk group. Such people need to keep their levels of bad cholesterol as low as possible, at around 1.8 mmol/L (70 mg/dL). In such cases, the margarine ads are right in saying we should eat as cholesterol-poor a diet as possible, with little fat and more dietary fibre. In addition, blood-vessel degeneration can be countered with regular exercise.

Unfortunately, however, not all patients' cholesterol levels can be adjusted in this way. The last resorts are cholesterol-dialysis therapy, in which excess cholesterol is mechanically removed from the blood, and medication known as statins, which inhibit cholesterol synthesis in the liver.

But what about eggs? Are they banished from the fridge forever? No, they're not. Rather than outlawing one foodstuff, it's more important to eat a rounded, well-considered diet. The Heart Foundation recommends — surprise, surprise! — eating a more Mediterranean-type diet, including lots of vegetables, salad, fruit, and wholemeal products. The American Heart Association even puts a number on it, saying that eating two eggs a week is harmless to health. But each individual is different, and the crucial point is how easily a person's body can break down cholesterol, which is mainly dictated by their genetic makeup. So, as a precaution, people with an existing cardiovascular condition should eat fewer eggs than others.

Anyway, the eggs the Easter Bunny brings these days are mainly of the chocolate variety. And I still love searching for them, and guzzling them down, as much as I did when I was a kid. That won't affect my cholesterol levels too badly, although it could hardly be called healthy, either. But, hey! The Easter Bunny comes only once a year!

Sweet by Nature

I grew up on a housing estate on the edge of a forest. My life back then consisted of building dens in the woods with my friends, or patrolling the neighbourhood on our bikes. We would invest our entire pocket money in sweets and cola from the kiosk on the edge of the woods and then stash our treasures in a secret hiding place. But our provisions never lasted long. Usually, we had scoffed them all by the end of the day, leaving us with both a sugar and caffeine rush.

Secretly snacking on sweets and sugary drinks wasn't a problem back then. At least, not for me. Perhaps more for my mother, who puzzled over why I had such trouble getting to sleep at night. One day, my friends and I were exploring the woods when we made an interesting find: an object that looked like a hypodermic needle. We didn't dare touch the mysterious article, since we'd had it drilled into us at school that we should leave such things alone and go fetch an adult! So off we rode on our bikes to a friend's house to tell his parents about our mysterious find.

The discovery of 'the needle' was the main topic of conversation on our estate for the next two weeks. Junkies in our perfect little world! A real small-town sensation. I was all the more astonished, then, to learn that what we had found wasn't a 'drug needle' at all. It turned out to belong to a local girl with diabetes. She had simply lost the syringe she always carried with her that contained her regular dose of insulin. This was my first ever exposure to diabetes as a child.

Until that time, I had no idea that sugar could have any harmful effects other than rotting your teeth and making you fat. Since I knew about those dangers — and because of my mother's constant nagging — I cleaned my teeth regularly. And I wasn't fat, either, since I was out all day, every day with my friends on our bikes. It was my mother who first told me about the other effects eating too much sugar can have on the body.

The term diabetes covers a number of metabolic diseases, which are all marked by the presence of sugar in the sufferer's urine. It's often noticed because patients get unusually thirsty and drink a lot of fluids. The most widely known form is diabetes mellitus, literally 'honey-like run-through'. This disease was described in documents as ancient as Egyptian papyri, although the name came about later, after physicians began to examine the colour, smell, and consistency of their patients' urine. And how did they identify the presence of sugar in urine in those days before laboratories and test strips? The answer is as simple as it is disgusting. Let me take this opportunity to offer up a vote of thanks to the pharmaceutical industry for inventing glucose test strips! As a doctor, the patients' wellbeing is of course paramount to me, but tasting a cup of their urine is definitely not number one on my to-do list ... In short, diabetes is called *mellitus* because it makes patients' urine taste sweet.

In order to understand how sugars are digested and

what the consequences of increased blood sugar on the cardiovascular system are, imagine you are a loaf of wholemeal bread, sitting on the dining table, about to be eaten. You're made up of protein, a very small amount of fat, some dietary fibre, and, predominantly, water and carbohydrates, i.e. sugar chains of varying lengths. No sooner are you in a diner's mouth than you are chewed up and drenched with saliva. Already, enzymes in that saliva begin to break down your sugar chains into their individual building blocks, especially disaccharides. This explains why bread begins to taste sweet if you chew it for long enough.

Slippery with saliva and properly pulped, you now slide through the throat, down the oesophagus, into the stomach and eventually the duodenum, the first section of the small intestine. Before your disaccharides can be absorbed here via the gut wall, they must be split into their two constituent parts. When we speak of blood-sugar levels, what we're really talking about is the blood's glucose content. That sugar is the body's most important energy source, because it's particularly quick at providing a boost. This should mean that a large amount of glucose in the blood makes us powerful and bursting with energy, but, as always, too much of a good thing is not healthy. An excess of glucose in the blood over a protracted period of time causes massive damage to our organs and blood vessels.

Urine begins to taste sweet when the blood's glucose content exceeds what doctors call the renal threshold. Under normal circumstances, urine contains no sugar at all, because the sugar is entirely reabsorbed by the kidneys before the urine reaches the bladder. However, this is only possible up to a certain level of sugar in the blood — the threshold value, which stands at about 10 millimoles per litre. Protracted high levels of glucose in the blood increase the risk of inflammations of the vessel walls and blockages in the tiniest of the arteries. Ideally, the concentration of glucose in the blood should be between 3.9 to 5.5 mmol/L before a meal, and 5.0 to 7.8 mmol/L after eating. That's equivalent to about a teaspoonful of sugar dissolved in around five to six litres of blood.*

In people whose sugar metabolism is working normally, blood-glucose concentration always fluctuates between those values, despite that fact that we take in huge amounts of sugar when we enjoy a nice piece of cake, and absolutely no sugar while we're asleep at night. In order to maintain this continuity, the body uses a simple trick: it stores glucose when levels are high, and releases those stores into the bloodstream when levels are low.

To do this, the body makes use of two important hormones: insulin and glucagon. They are the islanders among our hormones, because they are produced in isolated regions of the pancreas called the islets of Langerhans. Each of the two hormones has an opposite effect on blood-glucose levels. Insulin, whose name comes from the Latin word *insula*, meaning 'island', is produced by the islets' beta cells,

---

* Converted, the threshold value is 180 milligrams per decilitre. In this system of units, the limit values are 70–100 mg/dL before meals and 90–140 after.

from where it is released into the bloodstream. Its effect is to make our cells use up or store more glucose, or turn it into fat. That may not sound so great — after all, who wants more fat in or on their body? But without insulin, our blood-sugar levels would spike greatly after every rich meal and that would be far from good for our blood vessels. Insulin protects us from this by making sure glucose is removed from our bloodstream and stored in the liver in the form of glycogen.

So this is how insulin lowers blood-sugar levels — and it knows no limits. Left unchecked, it would continue to reduce the amount of sugar in the blood to the point of serious hypoglycaemia. This is where glucagon comes into play. Before blood-sugar levels fall to critical, glucagon throws a spanner in the works. As mentioned earlier, it is also produced in the islets of Langerhans in the pancreas, but this time by the alpha cells. Glucagon causes stored sugar to be released from the liver into the bloodstream, and, in emergency situations, even to be generated from scratch.

This is possible because our bodies are easily able to manufacture sugar from the products of muscle and protein metabolism. This process is so efficient, in fact, that it's theoretically possible to stay alive for a time without ingesting any external sugar at all. As adults, our bodies require around 200 grams of glucose a day, of which no less than 75 per cent, i.e. 150 grams, is used by the brain. Most of what remains goes to provide energy in our red blood cells. This means our blood-sugar levels should be dangerously low after a couple of days fasting on a desert island.

Luckily, however, that's not what happens. When blood-sugar levels fall below a critical threshold of around 3.5 mmol/L (60 mg/dL), our heart, brain, muscle, and, in particular, liver and renal-cortex cells are stimulated to make up for the missing glucose. The body is able to manufacture around 180–200 grams of sugar per day in this way. Which explains why our blood-sugar levels don't normally fall below 3.5 mmol/L.

This ingenious system seems almost too good to be true; and, indeed, it is rather vulnerable. Although our bodies are able to protect themselves from excessively high blood-sugar levels by secreting more insulin, this also leads to some rather unpleasant side effects. For example, increased insulin production can cause water and fat retention, high blood cholesterol, and, not least of all, high blood pressure. In addition, when insulin is regularly secreted in excessive amounts, our cells become increasingly insensitive to its effects; they become immune to insulin, as it were. Consequently, the next time we eat a high-sugar meal, our body has to secrete more insulin to gain the same effect, and a vicious circle has begun.

Carbohydrates are an essential part of a healthy diet, but this is only one side of the coin. Consumed to excess, they do more harm than good. There's no doubt that our brains and red blood cells cannot do without glucose, which is precisely why our bodies have come up with such a marvellous thing as gluconeogenesis (generation of glucose from the by-products of protein metabolism) over the course of evolution. For this reason, it's advisable to steer clear of chocolate cake, sugary drinks, and other sweet treats as much as possible, difficult as that may be.

Since so many people these days fail to take such well-meant advice to heart, they are increasingly prone to the consequences of excessive carbohydrate consumption. Thus, their blood pressure rises, their cholesterol levels reach dizzying heights, and abdominal fat — which, by the way, can be caused by extreme stress[*] just as much as regular boozing sessions — affects their appearance in a less-than-positive way.

When people consume too much carbohydrate and their cells become increasingly resistant to insulin, they are well on the way to developing type 2 diabetes, the kind which is acquired rather than congenital. It is often called adult-onset diabetes, which is misleading, since the number of young people developing type 2 diabetes due to obesity and lack of exercise is increasing drastically. This disease is rapidly taking on epidemic proportions: almost 90 per cent of diabetics worldwide have the type 2 form of the disease. And this doesn't include the estimated 200 million people who have already developed it, but haven't yet been diagnosed because their symptoms are still mild.

Diabetes is caused by the inability of the pancreas to produce the required amount of insulin, leading to a rise in blood-sugar levels. But where's the problem, when all sufferers need to do is pop a few pills, or inject a little insulin?

_____

[*]   See the chapter 'Sleeping Beauty's Heart', p. 249.

The sobering answer is that patients who inject insulin may see their blood-sugar levels fall, but their blood pressure, cholesterol levels, and percentage body fat will rise. This, in turn, increases the risk throughout the body of serious vascular damage: heart attacks, strokes, and — yes, even the unthinkable — erectile dysfunction all begin to loom on the horizon. In a considerable number of patients, entire limbs eventually become so cut off from a supply of blood that amputation is the only option.

The first amputation I ever witnessed firsthand in the operating theatre was that of the right lower leg of a diabetes patient. Actually, it wasn't really a complete lower leg anymore, since three toes had been removed three years earlier; and a year after that, half the patient's foot had been amputated. This vicious circle cannot be broken by injecting massive amounts of insulin — the only effective long-term solution is to follow a diet that is low in carbohydrates. This really is the best way to minimise the risk of diabetes-associated damage to the body. However, although most patients are completely aware of this, many find it extremely difficult to stick to such a diet.

Regular visits to the doctor are important to help patients maintain that necessary discipline. Together with their doctor, diabetics can draw up meal plans, seek out hidden sources of carbohydrate in their everyday habits, and, most importantly, confess their sugar sins of the previous week. Doctors will identify them anyway when they check their patients' blood-sugar levels. Yes, that demigod in the white coat knows everything!

A blood test of this kind will not only show the doctor a patient's current blood-sugar level, it will also show the levels of something called HbA_{1c}. This is a particular form of haemoglobin to which glucose attaches (and remains attached). Its concentration in the blood increases with increasing blood-glucose levels. The ratio of HbA_{1c} to normal haemoglobin in the blood provides a way to calculate the concentration of sugar in the blood four to 12 weeks in the past. Which will, of course, reveal any sugar sins. So, there can be no secret visits to the local snack bar after dark!

This leads patients to change their ways, and once they have experienced the initial success of their dietary discipline, many see their quality of life improve so much — despite the limitations on what they can eat — that they stick to their diets without their doctor's help.

Our body is like a cup of coffee. A little bit of sugar is fine, but too much ruins everything. And if other risk factors are also present, it can put our bodies, and in particular, our hearts, in great danger. If problems metabolising sugar are accompanied by high blood pressure, a lipoprotein imbalance, and 'central obesity' — medical jargon for a 'beer belly' — then we have a case of the 'deadly quartet'. The Bucks Fizz of risk factors, if you like. Also known as metabolic syndrome, these four things, along with smoking, are the most important risk factors for cardiovascular disease.

Metabolic syndrome is far more common in industrialised

countries than it is in less developed parts of the world. Why? It is simply down to our lifestyle, in particular our diet. Couch potatoes are most likely to suffer from metabolic syndrome, and both 'couch' and 'potato' are particularly apt words in this context. Carbohydrate-rich overeating together with extreme lack of exercise (as well as smoking) are without doubt the top causes of this condition.

Constant overeating leads to obesity and makes the body's cells resistant to insulin. Fat in the belly area is particularly dangerous as it also envelops the abdominal organs. When it breaks down, free fatty acids are released, among other things, which enter the bloodstream and cause the cells of our muscles and liver to be almost totally unresponsive to insulin. The result is a huge increase in blood sugar, and all the negative consequences this brings.

Be Still, My Racing Heart

The heart's electrical-conduction system, cardiac arrhythmia, resuscitation, and heart transplants

A Jackhammer in the Heart

One widespread medical condition that adversely affects the heart's ability to work as it should is cardiac arrhythmia — an irregular heartbeat. A healthy heart beats tirelessly at a steady pace, no matter whether we are awake or asleep, or whether we are exercising or at rest. Our heart never sleeps. If you take the time to listen carefully to your heart as it works, you can even feel your heartbeat slowing down when you are at rest again after climbing the stairs.

What happens in our chest when we take physical exercise is known to medics as tachycardia. That simply means rapid heartbeat; in this case, rapid is deemed to be more than 100 beats per minute. The opposite of tachycardia is bradycardia, which describes a heart rate of below 60 beats per minute. However, our heart can easily beat even more slowly than that. After we get up in the morning, for example. Even during the day, our heart rate — often also called the resting pulse rate — can be lower than 50 beats per minute. This is especially true for top-level athletes, whose well-trained hearts don't need to beat as fast when they are at rest in order to supply their bodies with sufficient blood and oxygen; the reverse is true for people who are less fit or whose hearts are weakened by disease.

Fitness level and physical exertion aren't the only factors

that influence our heart rate. It's also affected by our mood and emotional state, as I was able to observe on myself recently while stuck at a temporary roadworks traffic light on my way to an important meeting. During this forced wait, I had the time to tune in to my own heart, which was beating calmly away to itself at first. But then it got faster and faster as more and more red phases of the lights passed and I was only able to inch forward a tiny amount each time, with the hour of my appointment rapidly approaching. Eventually, my heart was pounding in my chest like a jackhammer.

But what if your heart begins to race like that for no obvious reason, when you are neither exercising nor under any kind of stress? Just like that, for ten minutes or so, while you are sitting on the couch? If this happens, there's usually a medical issue that needs to be investigated by a doctor. Spontaneous arrhythmias often turn out to be an indication of a serious heart problem, which can only be diagnosed by means of a thorough medical examination. Although, this can be a challenge for the doctor, since such brief arrhythmia episodes rarely occur in the doctor's waiting room or during an examination, making it difficult for a physician to assess them.

Nevertheless, it's important not to be deterred or ashamed, and to describe your symptoms as accurately as possible to the doctor. How often does your heart begin to race for no apparent reason? Does it always feel the same and last for a similar amount of time? Does it begin suddenly and quickly, or does your heart rate climb gradually? Does it stop again suddenly, or gradually, like a decrescendo at the opera? Does an episode last just a few minutes, or perhaps hours? And, one question that is particularly important in emergency situations: is your heartbeat rapid but regular, or is it both accelerated to more than 100 beats per minute *and* irregular (a condition called tachyarrhythmia), such as: ba-boom — pause — ba-boom, ba-boom — pause — ba-boom — pause — ba-boom, ba-boom?

It's extremely important for patients complaining of an irregular heartbeat to undergo an ECG, since symptoms such as shortness of breath and chest pains, along with arrhythmia, can be indicative of a heart attack. However, this does not mean that every time someone's heart beats a little irregularly, they should immediately be afraid they're having a heart attack. Although it's true that a heart attack is often associated with an irregular heartbeat, the opposite is not always the case, and an irregular heartbeat needn't always be an indicator of a heart attack. Even completely healthy people's hearts can occasionally stop momentarily, or feel as if they are beating irregularly in some other way.

It is only when such missed beats become a frequent occurrence, or when the heart starts beating wildly out of time, that doctors speak of cardiac arrhythmia. This does not by any means suggest that a patient is in immediate mortal danger, but cardiac arrhythmia can be bothersome in day-to-day life. Particularly when it adversely affects the body's ability to complete the most commonplace activities, like climbing the stairs or making the morning coffee. Sometimes, cardiac arrhythmia can cause dizziness or nausea. Whatever its effects, it is a nuisance. How much of a nuisance it can be was brought home to me during an emergency call-out close to where I live.

A sharp turn to the left, speed up on the straight, brake, a sharp turn to the right, accelerate, brake, another sharp right. The engine roars as we hit another straight section of the winding country road and my teammate Tom steps on the gas again.

'We're in an ambulance, not a fighter jet! Keep the speed down!' I hiss nervously. 'Otherwise, I won't be able to keep my breakfast down.' I'm smiling, but also clinging on to the grab handle for dear life, attempting to stop myself from being thrown from side to side in my seat. Tom seems to be enjoying this ride far more than I am. I've always known he's a fast, if extremely safe, driver, but this morning it's a little bit too much for me. Especially since we're only a few hundred metres from the call-out address: a terraced house where a 69-year-old male has suffered head injuries.

We cover the final stretch at walking pace, while I look for the house number. 'There it is,' I call, and my colleague is already parking the ambulance. An elderly man stands in the doorway, clutching a blood-soaked kitchen towel to his forehead. He invites us into the kitchen. A modern space, all brushed metal, with a solid wooden table in the middle. The polished kitchen tiles are flecked with blood. While Tom starts taking down the man's details, I take a look at his forehead, where I see a clean, two-centimetre-long gash. I quickly dress the wound.

The man tells us he blacked out from one second to the next and came round lying on the floor. On his way from the vertical to the horizontal, he must have hit his head on the edge of the table. My co-worker measures his blood pressure and pulse. 'Pulse easily palpable, normofrequent but arrhythmic,' he says, looking up at me expectantly. Arrhythmic! This means I should prepare an ECG. The electrodes are soon in place, and a line flickers across the screen. Tom and I watch it carefully, and both reach the same conclusion: arrhythmia, but no other abnormalities. We decide to take the patient to hospital without calling in an emergency doctor first.

Assessing such heart malfunctions is a delicate balance. Is the patient's condition stable or variable? Is his heart rate getting faster or slower, or is the ECG-reading constantly changing? And most importantly: how is the patient reacting? Can he be transported safely? In this particular case, the answer is already quite clear: the man is alert, his circulatory system is stable, and he's no longer bleeding uncontrollably. The only cause for concern is his cardiac arrhythmia, but this is also the answer to the puzzle. After excluding the possibility of an acute elevation infarction, we're pretty certain his arrhythmia was the cause of his sudden blackout. Sometimes, the interval between two heartbeats can be so great that the brain is briefly starved of oxygen. When this happens, the patient loses consciousness for a second or two, but has often already come round again by the time he hits the ground.

Such an episode is not usually life-threatening, unless the patient happens to be a tightrope-walker by profession. But coming round on the floor with a bleeding gash on your forehead is certainly a nuisance. That's why it is vital to consult a doctor after such an episode, to find out how to get the problem under control.

The most common arrhythmic event is one that I don't even really consider a medical condition. This is called a ventricular extrasystole, in which, rather than racing, the heart just stumbles a little. An extrasystole is an additional contraction of the heart muscles that doesn't fit into the normal beating rhythm. It's seldom associated with momentary blackouts. The worst subjective effect is that the next heartbeat following the extrasystole can feel like a gunshot in the chest, which may be extremely frightening. However, an extrasystole is no problem at all for an otherwise healthy individual. After all, it is really nothing but an extra heartbeat.

In such cases, it's sometimes advisable to take a 24-hour ECG reading, which means the 'patient' has to carry around a mobile ECG machine for a day. The resulting 24-hour cardiogram can provide clues for a more precise diagnosis. This may lead to a recommendation for a change in diet, a prescription for medication, or perhaps even a procedure called a catheter ablation to stop the heart from stumbling again.

A catheter ablation involves inserting a thin plastic tube (the catheter) into a blood vessel in the groin and advancing it as far as the heart. There, the damaged part of the heart muscle responsible for causing the arrhythmia can be accurately 'ablated' (the medical word for destroyed). Depending on which part of the heart is affected, a catheter ablation can be an extremely rapid procedure, or it can take several hours. Complications such as damage to the blood vessels or infection are possible, but exceedingly rare. The procedure is usually carried out under local anaesthetic and normally involves only an overnight stay in hospital.

According to the 2010 German Heart Report — an analysis of heart health, disease, and treatment published annually in Germany — 44,000 catheter ablations were performed in the country in that year. The procedure can be used to treat a number of different arrhythmia conditions, including atrial fibrillation, the uncontrolled twitching of the atria. One Spanish study showed that three out of every four patients who underwent the procedure were still 'twitch-free' one year later. Incidentally, such arrhythmia conditions are by no means restricted to elderly patients. They are also relatively common in younger people, both men and women.

Holiday-heart Syndrome: when a trip can make the heart stumble

A healthy human heart is like a well-oiled machine driven by various motors working together in perfect harmony. These motors glory in such names as the sinoatrial node, the atrioventricular node, the bundle of His, and the Purkinje fibres. Together, they form a kind of pacemaker, which produces electrical signals and transfers them to the heart muscle to make it work, i.e. beat. I'll explore the brilliant sophistication of this system a little later. First, I will examine the individual components of the heart's electrical-conduction system.

The chief pacemaker in this strictly hierarchical arrangement is the sinoatrial, or sinus, node. It dictates how often and how steadily the heart beats. Conditions such as high blood pressure, heart-valve defects, myocardial disease, or an overactive thyroid gland can cause the muscles of the atria of the heart to stop heeding the commands of the sinoatrial node. When this happens, the atrial muscles no longer work steadily and rhythmically, and their contraction and relaxation pattern starts to become uncoordinated. The result of this is atrial fibrillation, as mentioned earlier. However, this isn't the only serious consequence. The conduction of electrical impulses to the ventricles of the heart begins to go haywire, resulting not only in an unproductive twitching of the atria, but also an irregular pulse rate. When this happens, doctors speak of an 'absolute arrhythmia'.

Imagine your pulse is racing at more than 100 beats per minute even though you haven't done any physical exercise. You're having difficulty breathing, so much so that it becomes a real struggle for air, and you start to get frightened. Your chest feels constricted, sweat pours down your face — and then, from one second to the next, you feel absolutely fine again, as if nothing untoward had happened. What you just experienced is not a heart attack, and certainly not a figment of your imagination; it was atrial fibrillation. Emergency medics will take an ECG reading to confirm it beyond any doubt.

Atrial fibrillation does not in itself represent an acute danger for life or limb, but it can have dangerous consequences. The greatest danger is that blood can begin to swirl and eddy inside the atria, which can cause the blood to coagulate and form clots. These clots can be pumped out of the heart into the body's circulatory system and swept along with the circulating blood until they reach a small blood vessel whose diameter isn't large enough to allow the clot through. Thus, a clot, or thrombus, can act like a bung in the vessel, preventing parts of the body from receiving an adequate supply of blood. As with any thrombus, if this happens in the brain, the result is a stroke; in the coronary arteries, a heart attack; and in the pulmonary artery, a pulmonary embolism.

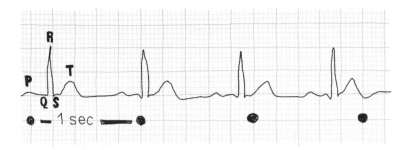

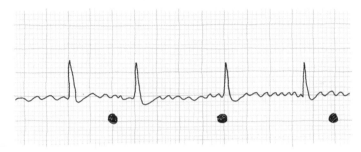

A sinus rhythm (top) indicates that the heart is working normally.
Atrial fibrillation (bottom) is clearly very different.

The increased heart rate caused by atrial fibrillation is almost like an endurance run for the heart, and, like a long-distance runner getting tired, the heart begins to weaken — until, in the worst-case scenario, it stops running altogether. If the fibrillation continues over several days or even weeks, the heart muscles will be considerably weakened. The result is heart failure, which was mentioned earlier.

Yet why should such a thing happen at all to a well-oiled machine like the heart? Especially since it's not restricted to older patients, but is also quite common among the relatively young? Apart from triggers such as a heart attack or insidious arteriosclerosis, there is one cause that is especially common among younger patients. It is a substance with many names. Chemists call it C_2H_5OH, but we know it better as booze, hooch, or firewater. Where I come from, we call it *Wegzehrung* — 'travel provisions'. This substance is, of course, alcohol. That's why atrial fibrillation is also known in medical circles as holiday-heart syndrome, since it often occurs following a bout of binge drinking at festivals or on vacation.

However, alcohol isn't the only risk factor for atrial fibrillation. Mitral-valve disease, various heart defects and inflammations, and even simple ageing are frequent causes, too. Although young people are affected relatively often, the risk of developing atrial fibrillation rises by almost 50 per cent every ten years for people over the age of 50. Those with high blood pressure are especially in danger. Their risk of developing atrial fibrillation is almost twice as high as that of people with normal blood pressure. Even conditions that would appear to have more to do with the lungs than the heart can cause this insidious fibrillation, including, for example, sleep apnoea — interruption of regular breathing during slumber.

Atrial fibrillation is one of the most common reasons for hospitalisation and visits to the doctor. The numbers are rising, which may be due, among other things, to our Western way of life. Happily, it's not only the number of cases that has increased, but also the number of treatment options. The life expectancy of atrial-fibrillation patients is increasing all the time. Those who are below the age of 65 can now expect to live about as long as people with no arrhythmia complaints.

One of the main reasons for this improvement is that such abnormalities are being identified and treated much earlier, and this is important because each acute fibrillation episode increases the likelihood that another will follow, until the condition becomes chronic. Practice makes perfect, they say, and in such cases, the heart practises fibrillating until it becomes an expert at it.

Atrial fibrillation that lasts longer than a week is described medically as 'persistent' and is characterised by the heart's inability to find its own normal rhythm again without help. Such help can take several forms. Patients can be prescribed medication that supports the heart's rhythm. Or doctors can use the brute-force method — better known in medical circles as electrical cardioversion — which involves jolting the heart back into a normal sinus rhythm with relatively weak but effective electric shocks.

If these methods fail to get the heartbeat back on track, the remaining options are limited. The main priority then is to deal with the high pulse rate. Doctors do what they can to reduce it to a reasonable level, but that doesn't remove the danger of clots forming in the blood. This problem can be tackled with blood-thinning drugs such as warfarin.

If drug therapy isn't effective in reducing the patient's heart rate as desired, one possible course of treatment is to destroy the atrioventricular node of the heart's electrical-conduction system, thus cutting the electrical connection between the atria and the ventricles. The problem with this, however, is that the chambers of the heart now have to be forced into rhythmic action with the aid of an artificial pacemaker. This is a pulse generator that replaces the natural conduction system. You can think of it as an extra spark plug in the engine of a car. If the heart's own spark plugs are no longer able to keep the engine going properly, a pacemaker will help to achieve this.

A pacemaker usually consists of a housing for a battery, and up to three wires, called electrodes or sensors. They are connected to the cardiac muscle, and monitor the heart's beating activity. When the heart slows down too much, or misses a beat, the pacemaker sends electrical signals to the heart's muscles to force it to expand and contract at the right tempo, so that the heartbeat becomes regular again. If the pacemaker didn't do so, the patient would be in danger of losing consciousness or suffering continuous dizziness. It's also not uncommon for the heart to beat normally while the patient is lounging around on the sofa, but to suddenly slow down greatly at the slightest amount of exertion. In such cases, a pacemaker can be a real blessing.

The first internal pacemaker was successfully implanted into a human in Sweden in 1958. Modern pacemakers are usually no bigger than 50 millimetres in diameter. Various models are available, depending on whether the pacemaker is a permanent or temporary measure, attached to the skin or inserted beneath it — so surgery is by no means always necessary. For some patients, it's enough to attach a large electrode to the skin above the heart, where it does its job by emitting regular electric impulses. However, these impulses need to be relatively strong, since they have to penetrate through the skin and all the way to the heart. This is the reason why such external — or 'non-invasive' — pacemakers are usually only used in emergency situations or when they need to take effect particularly quickly for some reason.

Another, albeit rather unpopular, way to stimulate the heart is via the oesophagus. Electrodes are passed into the oesophagus down as far as the region of the heart. From there, the heart can be paced using electrical impulses. Since it is a relatively unpleasant experience for the patient, however, oesophageal pacing is a rarely used technique.

It's also possible to place the pacemaker wire into a vein and pass it into the right side of the heart, with the actual pacemaker remaining outside the body. This kind of intra-cardiac stimulation can only ever be a short-term emergency solution, since any connection between the outside and the inside of the body is a possible gateway for disease-causing germs, making it a considerable target for infection. Germs should be kept out as much as is possible, so this must be done under sterile conditions.

When we talk about cardiac pacemakers, we usually mean a device implanted beneath the patient's skin or chest muscles. This sounds like a dramatic procedure, but in fact it's a relatively harmless operation for the patient and can often be carried out with just a local anaesthetic. A pacemaker of this kind can continue regulating the wearer's heart rhythm for around five to ten years.

But before such an aid becomes necessary, there are

many possible ways to protect the heart from arrhythmia. One very pleasant way is: relaxation. So, why not relax on a nice holiday? As long as it's not a binge-drinking rampage on the beach, the risk of developing holiday-heart syndrome is extremely small.

A Naturally Integrated Pacemaker

The muscles of our heart are made up of billions and billions of cells. These alternately contract and relax in a ceaseless rhythm, producing our heartbeat — which is actually nothing more than a muscular contraction triggered by an electrical impulse.

These impulses are created by a sophisticated system of specialised cells that generate the impulses and propagate them to muscles of the heart. They differ from other cells in that they can activate independently, without 'orders from above'. They basically behave like a workaholic who's happy to work overtime, without the need for a dressing down from the boss.

The chief impulse generator, or primary pacemaker, is the sinoatrial node, which I've already mentioned. This 'conductor' consists of a collection of specialised cells somewhere in the right atrium of the heart. If you asked me to describe the precise location of the sinoatrial node, I would be slightly challenged — almost like the satnav in your car when it gets confused. 'Turn around the superior vena cava into the right atrium. Your destination is on the left.' Looking left, I see nothing. 'Please make a U-turn.' I make a U-turn. 'Your destination is on the right.' Yeah, right where I just looked and saw nothing. Whatever. I play along, take a closer look and still don't see it. 'Please make a U-turn.'

My resentment towards the disembodied voice grows, but I continue to obey and keep searching. 'You have reached your destination!' Yeah, great! I still can't see a damn thing. But this time, for a change, it's not the satnav's fault. The sinoatrial node really is located near the entrance to the superior vena cava, but is all but indistinguishable from the surrounding tissue. The satnav can tell me which street it lives on, as it were, but has no idea of its house number.

The sinoatrial node works completely independently, with a frequency of 60–80 impulses per minute. The signals it produces travel to the atria, where they cause the muscles to contract. Even before this happens, the valves between the atria and the ventricles have opened to allow blood to flow through. The contraction of the atria now presses a few more millilitres of blood into the ventricles. When they're full, the valves re-close. At the same time, the sinoatrial node also sends a signal to the secondary pacemaker, the AV node. 'AV' stands for 'atrioventricular', describing its approximate location between the right atrium and the right ventricle. It's called the secondary pacemaker because it can generate electrical impulses completely independently, in a similar way to the sinoatrial node.

If the sinoatrial node should give up the ghost as a result of a heart attack, for example, the AV node can still generate 40–50 impulses per minute and so keep the heart pumping. As such, it could be seen as the back-up generator for the heartbeat. Under normal circumstances, however, it does not act independently, but rather functions simply as a conductor for signals from the sinoatrial node.

It doesn't do that immediately, but with a certain delay. This is called the atrioventricular-conduction delay, and it ensures that the muscles of the atria and ventricles don't contract simultaneously, but rather fire one after the other in quick succession. First, the atria contract and pump blood into the ventricles. Only then do the ventricles contract and pump the blood out into the circulatory system of the body or the lungs. Additionally, the AV node also functions as a kind of 'watchdog'. When necessary — that is, whenever it receives too many impulses — it can block them, just like a bouncer at the door of a trendy club. One time this happens is during atrial fibrillation.

Before the impulse from the sinoatrial node can travel via the AV node to the muscles of the heart, it passes through something called the bundle of His, which is located about one centimetre lower down, towards the apex of the heart (the perhaps-confusing anatomical name for the lower-left part of the heart). The bundle of His received its rather strange-sounding name from Wilhelm His, the Swiss cardiologist who discovered it. Like the sinoatrial and AV nodes, it can produce its own impulses if needs be. However, it has a much lower frequency than those two, of just 25–40 impulses per minute. If both nodes fall silent, the bundle of His can take on the responsibility of generating impulses as an emergency measure.

Luckily, in a healthy heart, this so-called ventricular escape beat never kicks in, since the normal signal from

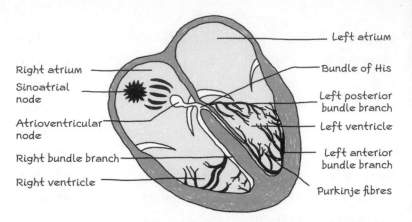

Right atrium

Sinoatrial node

Atrioventricular node

Right bundle branch

Right ventricle

Left atrium

Bundle of His

Left posterior bundle branch

Left ventricle

Left anterior bundle branch

Purkinje fibres

A general representation of the electrical-conduction system of the heart

the sinoatrial or AV node overlays the bundle's intrinsic frequency and it is only too happy to take orders 'from above'. The bundle of His is sometimes described as the tertiary pacemaker, the third impulse-forming structure of the heart's electrical-conduction system. The sinoatrial node is the big boss, so to speak, and the AV node and the bundle of His are the senior executives, issuing and passing on instructions, while others follow them.

Now we get to the level of the branch managers, as the bundle of His splits off into two branches, appropriately called bundle branches.* The electrical impulses pass from the bundle of His located in the cardiac septum (the dividing wall between the right and left sides of the heart) via the left and right bundle branches to a stringy network

---

* In fact, anatomists class them as part of the tertiary pacemaker, along with the bundle of His.

called the Purkinje fibres.* These fibres conduct the fine impulses to the muscles of the ventricles, which react by promptly contracting. Although this network is very intricately branched, it doesn't reach every single muscle cell. This explains why there are also electrical connections between the individual cells, like the synapses between nerve cells — but in this case, scientists refer to them by the uncharacteristically straightforward name of 'gap junctions',† and they conduct the tiny impulses to make sure these reach every part of the muscle tissue.

In a healthy heart, it's the sinoatrial node that dictates the pace of the heartbeat to the other parts of the electrical-conduction system. But why aren't the impulses transferred directly from the atria to the muscles of the ventricles, when even the most perfunctory of examinations will show that those two areas are located right next to each other? This is due to what anatomists call the cardiac skeleton, which forms a dividing wall of connective tissue between the muscles of the atria and the ventricles, preventing electrical impulses from passing through. After all, the pumping system depends on the atria contracting first, before the muscles of the ventricles follow with a slight delay. A pretty sophisticated mechanism, don't you think?

As well as being important that electrical impulses don't cross directly from the muscles of the atria to those of the

---

*   Also called the Purkinje tissue or subendocardial branches.

†   Also known as a nexus — or *macula communicans*, for those who prefer their medical terminology in Latin!

ventricles, it's also essential that the impulses behave in an orderly fashion within the chambers of the heart. If they were to start spinning around in circles within one chamber, or jumping from one side to the other and back again, the result would be horrendous. This would make it absolutely impossible for the heart to function regularly and provide the body with the blood it needs.

Luckily, a healthy heart can rely on a simple phenomenon to prevent this worst-case scenario from occurring. Once muscle cells have been stimulated, they need a short time to recover before they can react to stimulus again. (Any similarities to post-coital human males is purely coincidental ...) For a fraction of a second, these cells are completely unresponsive to any electrical impulses. If one should reach a cell during this short interval of time, the impulse simply dissipates without causing any damaging effects.

The heart's electrical-conduction system is a naturally integrated pacemaker. It is a marvellous machine. But even the best machines sometimes break down and need to be repaired. And, just like modern motor mechanics, doctors can use a diagnostic device to check whether the heart's conduction system is working properly or is in need of repair.

If You Can See the Steeple, the Graveyard Isn't Far

Often abbreviated to ECG,* the electrocardiogram is by far the most important diagnostic tool available to cardiologists and emergency medical staff. The word is made up of the Greek for 'electricity', 'heart', and 'write down', so it's a machine that writes down, or records, the electrical activity of the heart. That record takes the form of a graph line on a strip of paper or a screen. As I've explained, the heart muscles act in response to electrical impulses, and that electrical activity can be detected and recorded. To do this, electrodes are attached to the patient's chest to register the electrical activity of the cardiac muscle fibres, or, more accurately, the changes in voltage. Those measurements are then displayed as described above.

Medics use this technology in a variety of ways. The best-known use is to record a resting ECG — a procedure completed in a couple of minutes, which almost everyone will probably undergo at their family doctor's surgery at

---

* Some medical professionals, especially in North America, prefer to use the German abbreviation, EKG. Opinions differ over why; some say it is in recognition of pioneering work done in electrocardiography by German scientists, others say it is to safeguard against confusion with EEG — electroencephalogram, a procedure to monitor the electrical activity of the brain.

some time or other. This method is particularly popular in emergency medicine. However, the disadvantage with it is that the information on the screen (or print-out) only tells the practitioner what is going on inside the patient's chest at the moment of measuring — it is, so to speak, live (hopefully!).

If a patient experiences an irregular heartbeat during physical exertion but shows no symptoms or abnormalities while at rest, then a resting ECG is not the right diagnostic tool. In such cases, a medical practitioner will recommend a so-called stress test, or exercise ECG. This involves exposing the patient to physical exercise, in a sitting or semi-recumbent position, for about a quarter of an hour. Usually, patients are required to work with their legs in a kind of land-bound pedal boat or on a stationary exercise bike. The level of difficulty is increased gradually until the patient is too tired to continue, or continuing would put the patient in danger, although such a stress test yields better results the closer the patient gets to his or her limit of exertion. As a precaution, medical guidelines state that the test should be stopped when the patient's pulse rate reaches 220 beats per minute minus the patient's age in years. This means a rate of 150 beats per minute for a 70-year-old test subject, for example.

During the period of exercise, the patient's blood pressure and pulse rate are checked continuously, while an ECG is recorded and monitored for any changes, and the patient is monitored for any cardiac arrhythmia, chest pain, dizziness, or difficulty breathing. Blood pressure and pulse rate continue to be monitored during the patient's recovery phase as an indication of his or her fitness level. The faster the heart rate and blood pressure return to normal, the fitter the patient is.

Another, rather more extensive type of electrocardiogram is the long-term ECG. This involves the patient carrying or wearing an ECG device for one to three days to measure his or her heart activity constantly. This allows doctors to see whether the patient's heart rate is normal over the long term, and to spot any dangerous changes in that heart rate during everyday physical exertion. Such as sex. Or during a penalty shoot-out. Or at an Andre Rieu concert, when all that frenetic fiddling might send the subject into a frenzy.

A similar technology to a long-term ECG is called a cardiac event recorder, which patients also wear during their normal daily activities. It also makes use of electrodes taped to the chest and so on, but an event recorder only begins registering heart activity when the patient presses the 'on' switch. Otherwise, it records nothing.

Medics can use ECG recordings to identify heart problems because pathological changes to this organ have a measurable effect on its activity. If the electrical-conduction system and therefore also the heart's activity become disrupted as a result of a heart attack, this will be recognisable as a characteristic change in the line recorded by the ECG compared to that of a healthy heart. Spikes may appear in the wrong places, or not at all, or the distance between them might change, or, in the worst-case scenario

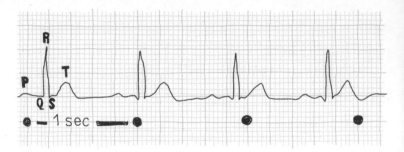

An ECG of a healthy heart (normal sinus rhythm)

of a cardiac arrest, the ECG shows nothing but the dreaded flat line.

The different spikes and waves recorded by an ECG are known by the letters P, Q, R, S, and T. Why they are not called A, B, C, D, and E is a mystery to me, to be honest. My guess is that the letters were inspired by the Cartesian coordinates system, invented by the French mathematician Descartes, who was in the habit of using a capital letter P to label a specific point on his graphs. Anyway, it doesn't really matter; what's important is that everyone knows what the various labels mean.

The P wave is created by the stimulation of the atria of the heart, i.e. by the impulses from the sinoatrial node and the resulting atrial excitation. This is followed by the largest feature on the ECG, which doctors call the QRS complex, with its two small downward blips surrounding one large spike. It's caused by the contraction of the ventricles, which begins with the Q wave and ends with the S wave.

That just leaves the T wave. It shows the repolarisation (or recovery) of the ventricles, and is an upward spike because

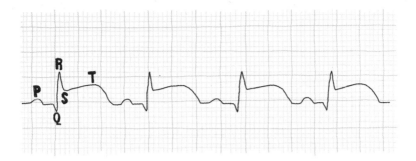

The elevation of the ST segment is a sign of a possible heart attack.
It looks like the silhouette of a church, complete with a steeple.

the process of repolarisation takes place in the opposite direction, from the apex of the heart to its base (the perhaps-confusing anatomical name for the upper part of the heart). In some cases, the T wave is followed by a so-called U wave (not seen in the illustration), which is usually caused by reverberating fluctuations from the repolarisation process.

Medical practitioners also divide the ECG trace into segments. The most interesting of those for emergency medicine is the ST segment. It has a significantly altered shape in the case of an oxygen deficiency due to a heart attack. When this happens, the ECG looks a little like a church with a steeple. A good way to remember this indication is the saying, 'If you can see the steeple, the graveyard isn't far.' A little macabre, I admit, but it certainly sticks in the memory. And this is useful when learning to interpret an ECG, which is a science unto itself.

One of the most common abnormalities recorded by an ECG is sinus tachycardia — an elevated heart rate of 100 beats per minute or above. This rate is normal in infants, but such a rapid heart rate in an adult is usually a sign that the heart isn't pumping out enough blood with each contraction, and is desperately trying to make up for this reduced blood supply to the rest of the body by beating frenziedly. This can be the result of a heavily bleeding wound or a shock, but it can also be caused by myocarditis (inflammation of the heart muscle) or heart failure. Even revulsion and fear, as instilled in me by far-right politicians, aren't enough to get the heart pumping properly. In this case, the coordinated pattern of the atrial and ventricular stimulation isn't actually out of synch; everything is just happing far too quickly.

And because anxiety or fear can raise the pulse rate considerably — cf. the abovementioned far-right politicians — patients often have elevated pulse rates when measured at the doctor's surgery, simply because they're anxious about having to see a doctor at all. A complicated conversation with the physician isn't likely to help relieve any of the anxieties of such patients. This is why it can be a good idea to gain some familiarity with the basic terminology surrounding ECGs before a consultation with a doctor about issues of the heart, so to speak. Doing so should help prevent any anxiety over medical mumbo-jumbo from making matters worse.

The opposite of tachycardia is sinus bradycardia — an abnormally slow heart rate of 60 beats per minute or fewer. As we saw earlier, this resting pulse rate can be normal for very fit athletes, but otherwise it may be due to an overdose of medication, an infarction, an underactive thyroid gland, hypothermia, or sick-sinus syndrome. The last covers a group

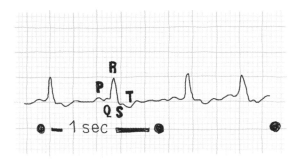

Sinus arrhythmia shows up as irregular intervals between QRS complexes

of heart-rhythm abnormalities caused by damage to the sinoatrial node, when the star conductor of the orchestra is stricken.

When the intervals between individual heartbeats as shown on an ECG are constantly changing, this is an indication of sinus arrhythmia — an irregular heartbeat rhythm. Simple breathing in and out can cause such an irregular heartbeat, but that's much more common in children and adolescents than in adults.

Much more striking is the abnormal sinus rhythm caused by atrial fibrillation — one of the most common forms of cardiac arrhythmia. As mentioned, this is when the atria start working completely chaotically. When one part of their musculature is contracting, another may be relaxing, and this shows up on an ECG as a lack of clear P and T waves surrounding the QRS complex, with the curve resembling a random scribble. It's easy to see why, since the stimulation is darting to and fro in the atria, where the muscles can't contract and relax in concert, but instead do

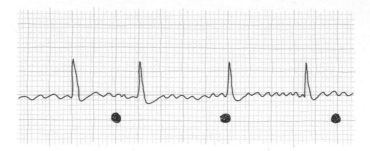

Atrial fibrillation: apart from the QRS complex, the curve is a chaotic 'scribble'

so without rhyme or reason. The spectrum of treatment for this condition, from blood-thinning medication to artificial pacemakers, should be familiar to you from the 'Holiday-heart Syndrome' section.

One particularly interesting disorder often encountered in the hospital or ambulance service is the atrioventricular block.* This occurs when the conduction of impulses from the atria to the ventricles is delayed or completely interrupted for a time. Depending on its severity, the condition is divided into three degrees. In first-degree atrioventricular block, the delay in impulse conduction from the atria to the ventricles is often so slight that the patient usually has no symptoms, and no treatment is necessary.

By contrast, in third-degree atrioventricular block, the impulse conduction from the atria to the ventricles is completely broken. In such cases, the answer is usually an artificial pacemaker! This is because, in the absence of impulses coming from the atria, the ventricles have only two choices: to hope that the pacemaker cells of the AV node —

---

* Or AV block.

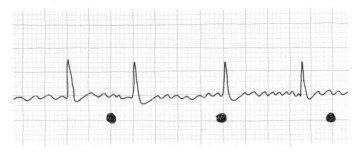

In third-degree AV block, atrial excitation (P wave) is disconnected from ventricular excitation (QRS complex)

the heart's secondary pacemaker — will kick in and provide the necessary impulses; or simply to stop beating. Even when the pacemaker cells do take over, the heart beats more slowly than usual. Too slowly to enable a patient to live well, and, unfortunately, this isn't a condition that gets better on its own without treatment.

The illustration above shows how the P waves reflecting atrial excitation are completely unconnected to the QRS complex of the ventricles' activity. This means there's been a fateful separation in the heart, the divorce papers are in the post, and the two sides are no longer talking to each other. My dear atrium, you'll be hearing from my lawyer!

The situation becomes critical if ventricular fibrillation occurs. Then, if no immediate help is at hand, death is the almost inevitable consequence, since ventricular fibrillation is the uncontrolled quivering of all the heart's muscles, which creates no palpable pulse. This means the heart is no longer pumping any blood around the body. Then it is a matter of mere seconds before the patient loses consciousness — and after a few minutes, lack of oxygen to the brain means it is about as active as a solar-powered pocket calculator during an eclipse.

The ECG is an indispensable tool for imaging what is going on in a patient's heart. But, as with all technical equipment, what appears on its screen should not be taken as an infallible oracle of truth. This is something I realised the first time I saw a doctor perform a precordial thump. This forceful blow against a patient's chest can be used to interrupt, temporarily at least, a life-threatening case of ventricular fibrillation.

On this particular occasion, it turned out that one of the sleeping patient's ECG electrodes had simply become detached. Although her heartbeat was still intact, the curve on the screen told a different story. The doctor should have used his fingers to check for a pulse in the time-honoured way, but he didn't want to waste any time, and whacked her square in the chest. This was not only extremely painful for the poor patient, but also so embarrassing for the doctor that, once he realised his mistake, he went down on bended knee to ask for her forgiveness.

Of course, the doctor had acted with the patient's best interests at heart, since drastic measures are necessary to get all the muscle cells of the heart working in concert again. The fact is that, in an emergency of this kind, those cells behave like schoolkids in an unsupervised classroom. They make nothing but mischief and continue to run riot until a teacher comes in and closes the door with a loud bang. Ideally, this will silence them all and they'll go back to concentrating on

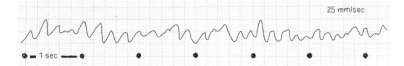

25 mm/sec

The muscles of the heart quiver uncontrollably, contracting and relaxing chaotically. Defibrillation and resuscitation are urgently required!

their work. In the case of ventricular fibrillation, the bang is created by a defibrillator, which uses a strong electric shock to stun the muscle cells of the heart into discontinuing their crazy, overexcited behaviour. This should put them in a position to obey the instructions from the heart's electrical-conduction system and begin working in an orderly fashion again.

Quit Playing Games with My Heart

Imagine you're walking down the street and, a few metres ahead, you see a man lying motionless in the dirt. You should hurry over to see if everything is alright and to find out if you can help in some way. Is it just a drunkard sleeping off his inebriation, or is some kind of medical condition responsible for the fact that he's snuggling up to the footpath? People suffering from hypoglycaemia (low blood sugar) are often mistaken for drunks. In either case, they need help from you as a passer-by. Perhaps they need an ambulance, or maybe they just need assistance getting vertical again. The only way to find out is to take heart and offer the person some help.

Unfortunately, however, the most common reaction when people find themselves in a situation like that is to cross over to the other side of the street as discreetly as possible, as if to say, 'What the eye doesn't see, the heart doesn't grieve over.' Why is this? To find out, students from the faculty of applied psychology at the University of Heidelberg carried out a detailed study of 'helping behaviour' among passers-by in emergency situations. The results were shocking.

The first test setting was a supermarket, where the students used hidden cameras to film people's reaction when the shopping bag of a nearby stranger burst, sending bread, fruit, tin cans, and yoghurt pots rolling all over the floor. Other students observed how many people were

168

prepared to help a wheelchair user board a local train. And in a third situation, they used actors pretending to be ill to study helping behaviour in medical emergencies. In one experiment, they planted their pretend patient on a bench in a pedestrian precinct. In another scenario, he was doubled up in apparent pain outside a train station. The researchers took care to make sure the situation was neither disgusting nor threatening for any potential helpers; in both cases, the situation was 'clearly recognisable as an emergency', according to the scientist in charge of the study.*

The students recorded how many people passed by without paying any heed to the person seemingly in need. Those who did spontaneously rush to the decoy's aid were then asked their age and, most importantly, about their motives for helping. The most common answer was that it was a natural reaction to offer help to those in need. But it appears it wasn't so natural for the majority of the passers-by. Some even complained loudly to the seemingly sick person that they should get out of the way.

The field study was continued over several weeks. In that time, a total of 94 people offered help to the person apparently in need. However, an incredible 6924 people passed by without doing anything. A shocking result! What causes people to simply ignore someone who is clearly in need of help?

Several theories have been put forward to explain this

---

* For example, the actors clutched at their bellies, groaned, and doubled over in apparent pain.

phenomenon. One of them places the blame on something called the bystander effect. According to this theory, the more people there are present to witness an emergency situation, the more likely individuals are to downplay its seriousness. I have witnessed this phenomenon myself at firsthand. Not so long ago, while travelling to Berlin with my mother, we had barely got off the train at the central station, when she grabbed me by the shoulder and cried out in shock, 'There's somebody lying on the floor!' I looked down the platform and, indeed, there was a man prostrate on the ground. He wasn't moving. And it was highly unlikely that he had chosen that precise spot to lie down for a rest.

At a guess, there were about 300 people in his immediate vicinity, probably more, and all of them were looking on with interest, but nobody was lifting a finger to help him. I really was the only person who did anything. When I bent over the man, who was clearly in need of help, I even heard someone behind me comment scornfully, 'He's just drunk!' Which was not completely wrong: the man was definitely drunk, but it was also clear that he'd fallen over and injured himself, and was in need of help.

As my mother and I were tending to the injured man, a woman approached and also offered to help. And then, suddenly, there was a tidal wave of helpfulness. This is typical in such situations. Until the moment when somebody finally takes the initiative, other bystanders will downplay the situation. This is due to a kind of self-reassurance effect, along the lines of 'If the situation were that bad, someone would surely already have stepped in to help.'

There's another phenomenon at play here, too, which everyone will probably recognise from personal experience: diffusion of responsibility. I spent a year sharing a flat with five other men in Vienna. Dirty dishes would often pile up in the kitchen, almost to the ceiling. None of us particularly liked having them there, but none of us felt responsible for doing something about it, either. We had subconsciously shared out the responsibility for the mess among us, and that is what psychologists call diffusion of responsibility. Everyone prefers to wait until someone makes the first move. Or in our case, until some six-legged little lodgers moved in, who certainly didn't chip in with the rent, and alien-looking landscapes began to flourish on the dirty plates.

A very common reason people have for not helping is simply that they are afraid of making the situation worse rather than better. They're afraid of failure or loss of face. I find this is understandable as, once again, I have personal experience of something similar. I remember my first ever resuscitation. I was 15, and it occurred some weeks after I'd finished my work experience in the hospital emergency department. It was early evening and I was at the railway station in Hanover, platform four. I was practically the only person waiting on the platform. Suddenly, I heard over the station's public-address system: 'If there is a doctor in the station, please make your way as quickly as possible to platform four!'

I wasn't a doctor, of course, but I felt somehow personally

addressed by the announcement nonetheless. Surreptitiously, I looked around. Sure enough, not 50 metres away, there was an elderly woman lying on the ground. She was lying face-up, completely motionless. *What should I do?* I thought. I was basically the only person who could help her, so I had to go over. My heart was pounding like a jackhammer, my legs had turned to jelly, and I was walking more and more slowly. *Dear God, please let a doctor turn up!* I looked round, but there was no one approaching.

I reached the woman and just stared at her for what must have been ten seconds, without doing anything. Her face was pale, almost like whitewash, her mouth was slightly open. Every couple of seconds, her lips would twitch, like those of a landed fish. I was completely helpless. I kept looking around in search of someone to help me. But there was still no one.

I pulled myself together, took a deep breath, and tried to remember what I had learnt in my first-aid classes. After all, I had practised dealing with precisely this situation, albeit on a plastic dummy. To lay hands on an actual real person now was probably the greatest challenge of my life so far. I tried speaking to her, shaking her shoulder. 'H ... hello, can you hear me?' No reaction. 'HELLO?' I shouted, now more emphatically, shaking her more forcefully. Nothing. She was unconscious. As I had been taught, I checked her breathing and pulse. No sounds of breathing; no detectable pulse.

Whew! Right then, here we go! I commenced mouth-to-nose respiration and cardiac massage. CRACK! The first rib was broken. I apologised to the unconscious woman and continued the procedure. After four cycles, I checked her breathing and pulse again. Still nothing. *Keep going.* And there it was again. CRACK! Another broken rib. This time, I didn't bother to say sorry.

During the sixth cycle, a man sauntered up to me, eating ice-cream with a spoon. 'Hello, I'm a doctor. What do we have here, then?' he asked calmly.

'What does it look like?' I snapped at him, completely stressed. 'Please help me!'

With a slow, deliberate nod, he set his cup of ice-cream on the ground.

'Could you take over the respiration?' I asked him. He nodded silently. We continued to resuscitate the woman for a few minutes, which seemed like hours to me. After what was probably about 20 cycles, I finally heard an ambulance siren. And suddenly, I could feel a pulse. And the woman's breathing had returned — shallow, but definitely audible. We carefully placed the old lady in the recovery position just as the paramedics arrived.

I have never since felt the fear of failure as sharply as I did on that evening. And I have never forgotten how petrified, indeed, overwhelmed I was by it. Today, I know that feelings of fear and self-doubt are normal in unfamiliar, stressful situations. And the only way to get over them is to face up to the challenge. When in doubt, it's always better to do something that might not be quite right than to do nothing at all. Leaving someone to their own devices when they are in need of help is just not an option! Especially if they have suffered a heart attack. After all, what outcome can be worse than death? A couple of broken ribs during cardiac massage? Certainly not. Older people's ribs break especially easily. It's an injury you just have to accept in such cases.

So remember: if someone needs your help, muster all the courage you have and help them! And don't hesitate!

There's a number that can help you in such situations: 110. No, that's not a new emergency-services number to call. It's a musical beat that can be very helpful for people who aren't trained in resuscitation procedures. Since my experience on the station platform, the American Heart Association has changed its guidelines for what it calls 'lay rescuers', so let's go through them. Imagine you're in the same situation as the 15-year-old me on that station platform. You approach the person who needs your help — hopefully more confidently than I did. The first thing you should check is whether the person is responsive and coherent. If so, a simple conversation is often enough to find out what needs to be done next. Patients are often able to explain what the problem is themselves.

If the person is unconscious and showing no signs of life, the first thing to do is to make sure the emergency medical services are on their way as quickly as possible. These days, if you are in a public place, there will often be an automated external defibrillator (AED) designed for use by non-professionals installed somewhere nearby. If so, ask someone to fetch it. While that's happening, check the patient's breathing. The best way to do this is to kneel down by your patient's side and place your ear above his or her mouth and nose with your head pointing towards the patient's feet. In this way, you'll be able to feel any breath on your cheek and at the same time watch whether the patient's chest is rising and falling.

If you detect breathing, carefully place the unconscious person in the recovery position. First-aid courses used to train participants how to do this in five steps, but today the three-step method is usually preferred. In my experience, these multi-step procedures only serve to confuse lay

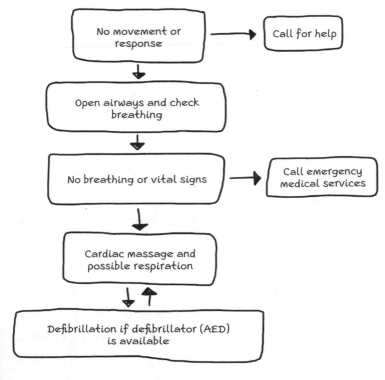

Resuscitation procedure

rescuers. I think it is better simply to remember the purpose of the recovery position: to make sure the patient is able to breathe freely. That's the only explanation you need. People often vomit when they're unconscious; when they're in the recovery position, the vomit can simply drain out of their mouths, rather than running back down their throats and blocking their airways.

If you do not detect any breathing, you must begin resuscitation straight away. Now is the time when it would be useful to have a defibrillator at hand. These devices are designed to be used with absolutely no prior experience. They give precise instructions for use, either optically via a display screen, or verbally via a speaker, and analyse whether the patient needs to receive a shock or resuscitation procedures. If no such device is available, then you must rise to the challenge yourself.

There's no need to check the patient's pulse again, since, if he or she is no longer breathing, you can reasonably assume that their heart has also stopped. According to the latest guidelines published by the American Heart Association, respiration is no longer a necessary part of resuscitation measures. That's good news. After all, who relishes the idea of placing their mouth on the face of a person who may well be dirty and is almost certainly a complete stranger? Research has shown that continuous chest compressions are much more effective in restarting, and then maintaining, blood circulation than a combination of interrupted compressions and the relatively unimportant 'kiss of life'.

The oxygen reserves in the patient's blood are in fact sufficient to last for the few minutes before trained medical personnel usually arrive. And anyway, performing cardiac massage is already enough of a challenge for someone who has never done it before. But how is it properly performed?

First of all, expose the chest. Not yours, of course, but that of the unfortunate individual lying helplessly in front of you. Open their jacket or shirt, or lift up their sweater or t-shirt until you see their breastbone. Then identify the compression point. For most people it is in the middle of the chest, halfway between the nipples. But what do you do if the person needing your help is a rather more pendulous person, for whom 'between the nipples' is somewhere in the region of his or her belly button? In that case, use your fingers to feel for the tip of the breastbone (where the ribs come together). Then, if you are right-handed, place the index and middle fingers of your left hand over the tip of the breastbone and place the ball of your right hand on top of them. For left-handed people, it's the other way around, of course.

Now it's all about the beat — the speed at which you should push down repeatedly (called compressions). And this is where the number 110 comes in. The optimum number of compressions per minute is between 100 and 120. This is a pretty rapid pace and not so easy to maintain over a long period. But, as coincidence would have it, there's a great way to maintain the right tempo and keep the ultimate goal in your mind: the required rhythm is the same as that in the appropriately named Bee Gees hit 'Stayin' Alive'. Another pop song with the right number of beats per minute and a similarly appropriate title is 'Quit Playing Games (with My Heart)' by the Backstreet Boys. Another one — my personal favourite, perhaps due to its rather less appropriate title — is 'Highway to Hell' by AC/DC. Those with more traditional tastes in music might like to compress along to the rhythm of Strauss's *Radetzky March*, which also has the right number of beats per minute. It helps to sing along to your chosen hit as you resuscitate, but only in your head, lest you offend bystanders. Especially if your song of choice is 'Highway to Hell'!

The next consideration is how hard to press. Resuscitating someone is hard work. Even if no ribs get broken, it can be a backbreaking job to take on the work of a person's heart over a protracted period of time. It's amazing to think that the little ticker in your chest normally manages the job all alone, isn't it? To make sure the heart gets compressed through the patient's skin and bone, you need to push the breastbone three to five centimetres into the chest with each compression. If you are doing it right, the patient's face will soon lose its pallor and gradually regain a rosy complexion. However, if you break five ribs with your first three compressions, the likelihood is that you're doing it too hard. The optimum amount of pressure varies from person to person. As a rule of thumb, you'll need to push less hard for someone with a delicate frame than for a 150-kilo body builder.

And finally, the most important thing: do not stop the

cardiac massage until another person is available to take over from you (or unless you'd be putting yourself in imminent danger by not doing so). In my work as a paramedic, I often saw people giving up on cardiac massage as soon as they heard the ambulance approaching. This can seriously jeopardise any success you may have had until then. Always keep compressing until someone takes over!

The best way to practice is at a first-aid class. Courses are held regularly and aren't expensive. And, as with everything else, practice makes perfect for lay rescuers. The first-aid course you reluctantly attended all those years ago is no help to you now. I know from my own experience that those with up-to-date first-aid training have more confidence in all walks of life, simply because they don't have that fear of failing in emergency situations.

I recently spoke to a boy of primary-school age who had come upon his father lying unconscious in the living room at home. Without a moment's hesitation, he called the emergency services and then ran to fetch a doctor whose practice was luckily just down the road. After that, everything happened quickly: the ambulance arrived, followed shortly by a rescue helicopter, which rushed the man to hospital. It later turned out that the man had suffered a severe stroke. Without his little boy's plucky intervention, the man would undoubtedly have died. What a hero!

Engine Trouble

Any car's engine will break down eventually. If the body of the car is still okay, the garage can replace the engine with a new one. A similar thing happens when a person's heart starts to perform increasingly badly and eventually breaks down altogether. The only differences being that the replacement work is carried out in an operating theatre rather than a garage, and is done by a heart surgeon, not a motor mechanic. Such an operation, better known as a heart transplant, was first performed successfully in 1969 and has now become a routine procedure. More than 5000 heart transplants are carried out worldwide each year. Under normal circumstances, the replacement heart, just like a new car engine, then performs perfectly.

But who can get hold of such a donor heart, and how? You can't just order one from a dealer, after all. Well, doctors can, in a way, but unfortunately donor hearts are a very rare commodity and demand far outstrips supply. That's why not everyone with an unhealthy heart can expect to receive a donor organ immediately. Furthermore, not just any old heart will do. The donor heart has to be matched to the recipient's blood group and body size — weight and height must not differ by more than 15 per cent, for example.

Patients are placed on the waiting list for a donor heart only if they fulfil a number of criteria. The most important one is that the patient is in urgent need of such a transplant. This may be the case if medication is no longer effective, and further increasing the dose isn't possible, or if all previous treatments have failed. Or if other operations on the heart, such as bypass surgery,* placement of a stent,† or heart-valve repair surgery are not possible or not advisable for some reason. Since so many candidates fulfil these criteria, patients can often spend years waiting for a donor heart. Patients who can no longer wait without a high risk of death are sometimes fitted with an artificial heart as a temporary measure until a real living organ becomes available.

In this context, 'temporary' can be quite a flexible term, however, since modern artificial hearts are designed for use over longer periods of time. This was not always the case. Early artificial heart devices were far more susceptible to the effects of wear and tear, which was partly due to the fact that they were built to imitate the pumping action of the entire heart. Today it's usual to make use of simpler devices that only assume the work of the left ventricle. These are known as left-ventricular assist devices, or LVADs for short. They are usually used in conjunction with blood-thinning medication, which helps the patient's blood flow more easily.

For patients with particularly erratic heartbeats, it's often enough to implant a small defibrillator, which automatically

---

* A procedure to restore blood flow to the heart muscle by diverting the flow of blood around a section of blocked coronary artery.

† A small, expandable tube inserted into a blood vessel to support it and keep it open.

delivers an electric shock (larger than that of a pacemaker) to the heart to return its rhythm to normal when necessary.

Anyone lucky enough to have received a donor heart has reason to rejoice and should look after their new ticker carefully in the course of their 'second life'. There's no shortage of romantic-comedy plotlines in which the recipient of a donor heart becomes the great love of the donor's mourning widow. Such films may not be particularly realistic, but they do show the way a whole new life opens up for heart-transplant patients. They can experience a real fresh start, with delicious Mediterranean-style food to enjoy, the prospect of many extra years of love, pleasure, and happiness, and a living, constantly beating heart.

Bedroom Sport for the Heart

A strong immune system, a lot of sex, and what that has to do with a healthy heart

The Sinful Way to a Healthier Heart

A candle-lit room; Marvin Gaye's 'Let's Get It On' is playing in the background. The curtains are drawn. Two empty wine glasses stand abandoned on the table. A schmaltzy still-life, innocently arranged and embellished with a number of hastily removed items of clothing. There, a pair of trousers; here, a shirt; and a few feet away, a pair of black briefs. Hopefully, a pair of socks also lies strewn about somewhere.

This trail of clothes leads to a couple engaged in an intensive workout. What they're doing is not only a lot of fun, but also good for their hearts, their immune systems, their general wellbeing, and their fitness. An entire-body cardiac workout. Most people don't exactly associate physical exercise with the idea of fun. When they think of sport, they think of sweat and slog, pounding the pavement every day through wind and rain, and immediately lose any spark of motivation they might have had.

It can be a challenge to break down that mental barricade. But I think I've found an alternative — and it's called sex! Of course, this can also bring you out in a sweat, but almost everyone in the world is wild about it nonetheless. And the best thing is: every time we jump in the sack together, we are doing our health a huge favour. This means, the more often you do it, the better!

Frequent sex is in fact an excellent way to combine

stress-reducing effects with physical exercise, and to have a lot of fun in the process. Furthermore, the hormones produced by our bodies when we have sex protect us from all sorts of illnesses and diseases. Indeed, one academic study showed sexually active people have a significantly lower risk of suffering a heart attack than the more sexually abstemious. Those hormones begin to flow even at the first tender touch, and increase to become a real hormonal fireworks display when we reach orgasm, which sends more than 50 different chemical messenger substances coursing through our veins. So, let's take a look at some of the most important ones.

Oxytocin: the cuddle hormone

Oxytocin is one of the most fascinating substances our body's chemical factory has to offer. It is not only produced by women as their bodies prepare for childbirth and during breastfeeding, in which instances it was first discovered; its release is also triggered by feelings of love. This has led to its becoming known as the cuddle hormone or the love drug. Once it is released into the bloodstream, oxytocin attaches to special receptors in the walls of various cells, depending on what kind of tissue those cells are part of. This means the hormone can have a range of different effects.

The hormone's name comes from the Greek *oxytokos* meaning something like 'swift childbirth'. In pregnant women preparing to give birth, oxytocin causes the muscles of the uterus to contract convulsively — a painful part of the birthing process. This effect explains why it is also used as a medication to artificially induce or speed up a woman's labour. If the hormone attaches to the cells in the glands of a nursing mother's breast, the glands are stimulated to produce milk. When a mother then breastfeeds her baby, the hormone has a relaxing effect on both her and the child, helping to cement the emotional bond between them. This is reinforced by the fact that the act of suckling also stimulates the baby's body to produce its own cuddle hormone.

Oxytocin may also be instrumental in forging the bond between partners in a monogamous relationship — although scientists have so far only managed to prove this effect in rodents, or, more precisely, in prairie voles. These little creatures change sexual partners very rarely, in contrast to their close but more promiscuous cousins, the montane voles, who are totally shameless in this respect. Researchers noticed that prairie voles not only have significantly more oxytocin in their blood than their more wanton cousins, but also have a different distribution of oxytocin receptors in their bodies. This led the scientists to suspect that the hormone could be responsible for the faithful behaviour of these creatures towards their sexual partners. To test this, they gave prairie voles an oxytocin antagonist, which inhibited the hormone's effects. And without so much as a by-your-leave, these otherwise thoughtful rodents upped and left their long-term partners and merrily began mating with all and sundry. To put it plainly: they were no longer interested in being faithful to their partners.

But can those experimental results be applied to human behaviour? To find out, a Zurich-based researcher carried out a much-publicised experiment that involved 49 volunteers playing a Monopoly-type game with clearly defined roles — some as investors, others as 'trustees' vying for the investors' money. Half of the investor group were given a nose spray containing oxytocin, while the other half were given a placebo with no active ingredients. Without so much as a by-your-leave, the group who were given the oxytocin spray began merrily mating with each other ... Only kidding — that didn't happen, of course. But what did happen was that those who received oxytocin were far more willing to hand over more of their investment money to the trustees than the control group. However, the effect only manifested itself if the two sides held face-to-face negotiations. If the investment discussions took place via computer, therefore anonymously, there was no difference in the trust behaviour of the oxytocin-doped and the non-oxytocin-doped groups.

It seems that in us humans, the cuddle hormone increases our inclination to trust others and thus improves our social skills, so to speak. It seems that it really does turn us into nicer people. And healthier ones, too, since oxytocin has also been proven to promote the healing of wounds and to lower blood pressure. It's a great all-rounder, known to have a calming effect and reduce stress. Ideally as a result of good sex.

Dopamine: the reward drug

Ahh, this is the life! I'm sitting with a cold beer in hand and a cigarette between my lips, next to a glowing barbecue, enjoying the enticing aroma of steaks sizzling on the grill. The sun is shining down from a cloudless sky, a pleasant breeze caresses my skin, and I'm totally content and relaxed. This is how a holiday should be. Soon, I'll be tucking into some tasty, if not particularly healthy, food. And the beer in my hand will certainly not be my last this evening. But, hang on ... Why is it that alcohol, cigarettes, and unhealthy food make us feel so good? After all, they're harmful to us, and we know it.

It's at moments like this that I realise that my body is that of a wild animal. I am perfectly aware of the fact that I'm in the process of doing my body a great disservice, but that knowledge does nothing to stop my body from gratefully accepting the rewards I'm giving it. The reason for this is that, at such times, unbeknownst to us, certain glands are pumping out a very special substance: dopamine, generally known as our reward hormone. It courses through us whenever we treat ourselves, whether that be by biting into a fresh, crisp apple, going out to engage in a bit of 'retail therapy', or lighting up a cigarette. Whatever it is that we happen to derive pleasure from.

When our bodies produce dopamine, we experience a pleasant feeling of reward and a warm feeling of contentment. What a shame that so many of the things that trigger these feelings are so bad for our health. All the better, then, that having sex — and, in particular, having an orgasm — causes dopamine to flood our system. This is useful, since all that fun under the bedcovers is what keeps our species alive. Or to put it another way: dopamine saves us from extinction. A pretty clever trick on the part of evolution. Contrariwise, too little dopamine flowing through the system can be a factor in causing depression.

As well as dopamine, another hormone, called noradrenaline (also known as norepinephrine, especially in the United States), is a significant factor in producing feelings of happiness. It's a precursor of adrenaline (epinephrine) and is manufactured by the body out of dopamine. When it is released into the blood in large amounts by the adrenal glands, we find it easier to concentrate; we are less susceptible to fatigue and hunger, and less sensitive to pain. It is often said of psychopaths that they constantly seek the rewards of a dopamine rush. According to this theory, love and mental illness aren't so far removed from each other, biochemically speaking. Not really surprising, somehow.

Adrenaline: the stimulator

There stands that person, the object of your desires. There's chemistry between you, and you're totally energised. Your heart begins to beat harder and faster, you are charged with energy, nothing and no one can stop you now. The cause of all these feelings is adrenaline, a stress hormone produced in the adrenal glands, which triggers a fight-or-flight response in us within a fraction of a second.

189

When we're being chased by a hungry lion — thankfully not an all-too-common situation to find ourselves in — it is adrenaline that makes us run like Forrest Gump. And when we're under attack, adrenaline gives us a feeling of power and determination the like of which we would never have thought possible in normal circumstances. Under the influence of adrenaline, a person being pursued is suddenly able to shift huge boulders that they would normally be incapable of budging even an inch. They can sprint distances only half of which would normally leave them half-dead and panting on the ground.

Funnily enough, something similar happens when we encounter our cherished heartthrob, or even when we just think of him or her. Adrenaline causes the bronchial tubes to expand so that we can breathe better; it causes the pupils to widen so that we can see better; it increases our breathing rate and our blood pressure; and from one second to the next, it causes a healthy heart to beat harder and faster — during sex up to a rate of 120 beats per minute. For our cardiovascular system, this is like a fitness machine inside our own body. Moreover, this stimulant increases our blood-sugar level, giving us more energy.

Unfortunately, all this can represent a danger for a weak heart, since adrenaline can place huge demands on it in stressful situations. Adrenaline is used in emergency medicine, but an overdose can disturb the blood flow in the heart, causing heart failure, a heart attack, or sudden cardiac arrest. Despite this, many people are so euphoric following an adrenaline rush that they become addicted to the hormone. Such people often become involved in extreme sports, taking ever more reckless risks in search of the ultimate adrenaline kick.

Serotonin: the happy hormone

Serotonin is the supreme happy hormone. It makes us feel pleasantly relaxed and content. It also bolsters our immune system, strengthening our body's defences against disease. When we are under the influence of serotonin, we feel peaceable and tend to see the world 'through rose-tinted glasses'. Doctors make use of this effect to treat depression, which can often be caused by depleted serotonin levels. When this is thought to be the case, a doctor can prescribe drugs called selective serotonin re-uptake inhibitors (SSRIs), which limit the body's ability to reabsorb the serotonin it produces. Serotonin is also instrumental in producing the feelings of happiness associated with sex. And it helps wounds to heal by causing smaller blood vessels to contract, reducing blood loss. So being happy isn't just positive in itself, it also makes us healthier. If that's not a win-win situation, I don't know what is!

Testosterone: the power source

Testosterone is one of the most important sex hormones, because it increases sexual arousal and the libido. A certain amount of the hormone is always present in the bloodstream of men and (to a lesser extent) women, and it decides whether and how quickly we can become sexually aroused. The more stressed a person is, the more difficult it is to 'get it on', in the words of Marvin Gaye (see earlier). This is because stress causes the bloodstream to be flooded with cortisol, the great hormonal antagonist of testosterone. However, if testosterone concentration in the blood gains the upper hand, a cycle of sexual stimulation begins.

This cycle occurs because testosterone regulates its own production. When there's an abundance of this lust-promoting substance, the pituitary gland sends out signals to boost testosterone production even further. In men, testosterone is produced by cells in the testes called the Leydig cells; in women, the hormone is produced by the theca cells in the ovaries. Women have less testosterone in their system than men because some of the testosterone they produce is converted into the female sex hormone oestrogen, which is responsible among other things for breast development, preventing bone loss, and increasing the concentration of good HDL cholesterol in the blood.* In men, testosterone promotes muscle growth, helps burn fat, and lowers cholesterol levels in the blood, which, as we have already seen, helps protect against vascular disease.

---

* See also, 'Should the Easter Bunny Go Vegan?', p. 118.

Endorphin: the painkiller

Endorphin could be described as the junkie among the hormones released during sex. Even its name makes this clear, being made up from the words 'endogenous morphine' (that is, morphine produced within our own body). Endorphin is well known to be an extremely effective painkiller. It inhibits the transmission of pain signals and helps us sleep better. Our bodies produce it in great quantities whenever we laugh, eat something delicious, or engage in strenuous physical exercise.* And, of course, when we have sex. Which goes some way towards explaining why people — in particular, men — tend to drift happily off to sleep soon after the act is done.

Oestrogen: the lust hormone

The most important female sex hormone is oestrogen. If we want to be precise, we should really talk of oestrogens, in the plural, because there are many subtypes of the hormone, although they all have similar effects on the body. One of the main influences they have on women is to cause them to have a greater appetite for sex just after ovulation — that is, when they are at their most fertile. This is, of course, another of evolution's canny tricks for keeping the species alive.

Researchers who analysed data from the Women's Health Initiative† found that in post-menopausal women — those

---

* See also, 'Jump, Heart, Jump', p. 214.

† A set of clinical studies to investigate the health problems of older women.

who are no longer menstruating — oestrogens even have an influence on the health of their joints and can inhibit joint pain. The Endocrine Society (an international medical organisation in the field of endocrinology — the study of hormones and the glands that produce them) suspects that this welcome effect is due to the anti-inflammatory properties of oestrogens. Furthermore, these hormones also appear to promote wound healing. In more detail, this effect is due to the fact that oestrogens inhibit the body's release of the cytokine MIF,* a substance that attracts inflammatory cells in their masses. It works in the same way as gentrification: one annoying yuppie moves into your peaceful little neighbourhood and starts showing off to all his old schoolmates about how great it is. And, before you know it, the area is full of people like him and you can no longer afford your rent, or even a mocha frappuccino from the hip new cafe on the corner.

What you need now is the anti-gentrification police: oestrogen. When it restricts the amount of the cytokine MIF, an inflammation is much less severe. Luckily, however, this doesn't mean women need to take medication to maintain higher oestrogen levels. That brings its own health hazards, including, for instance, an increased risk of breast cancer. A much better option than medication, and certainly a more enjoyable one, is to have as much sex as possible.

---

* Macrophage-migration inhibitory factor.

194

Bedroom sport provides a great way to combine physical exertion with stress-reducing effects while protecting our bodies by means of the hormones that sexual intercourse releases inside us. This hormonal cocktail is even more effective if we actually love our sexual partner. For example, sex without genuine affection triggers far lower levels of oxytocin release. The best advice, then, is not only to have sex, but also to 'make love' in the truest sense of the phrase.

Yet, like any medication, sex also brings with it certain risks and side-effects. Notwithstanding the kind of freak accidents you might see on reality TV shows like *Sex Sent Me to the ER*, vigorous sexual activity can be counterproductive to the health of those with pre-existing cardiovascular conditions. Thus, the most common cause of death during love play is a stroke or cerebral haemorrhage. This is because really 'going for it' with a sex partner, so to speak, causes a rise in blood pressure, which can be too much for already-damaged blood vessels in the brain to withstand. So sex is the best medicine for preventing cardiovascular disease, but it isn't effective as a cure. However, our bedroom antics certainly do boost our immune system and this lowers the risk of inflammation throughout the entire body.

Well, what are you waiting for? Go to bed!

The Body's (Almost) Invincible Army

No matter where we go, little critters are lying in wait in every nook and cranny. They hide out on door handles, computer keyboards, staircase bannisters, and even our own skin. I'm talking about those tiniest of creatures: single-celled organisms, bacteria, viruses, and fungi — which can all cause disease. Although, to be fair to the bacteria, I must say that the vast majority are perfectly harmless for human beings.

We have our immune system to thank for the fact that our body and its organs aren't overrun by these microorganisms; indeed, many of them even live in symbiosis with us.* It's our immune system that is responsible for maintaining a happy balance between our needs and those of our tiny companions. This heroic feat generally goes unnoticed by us, at least until our system is forced to try to restore that balance after it has been thrown out of kilter by illness.

If our body's army — a pretty accurate way of describing our immune system — did not constantly and aggressively rage against every pathogen we encounter, we would soon succumb to myocarditis† or any number of other illnesses. However, it doesn't only attack troublemakers from outside our body. It also fights against the body's own cells if they become degenerate or malignant. This is a good thing, of

---

*   Two different organisms living together to the benefit of both.

†   See also, 'A Red Card for the Heart', p. 209.

course, as someone has to maintain peace and order in the house. Sometimes, however, our body's defence system overreacts and begins attacking healthy structures, causing major problems in the house when someone is just listening to loud music or hasn't tidied up recently. When this happens, doctors speak of autoimmune disease.

Biologists divide the immune system into two components: the innate and adaptive (or acquired) immune response. Both components have the same basic task of preventing pathogens from spreading through our bodies and attacking our tissues and organs. But there are major differences between the innate and the adaptive immune systems.

The innate immune system

As the name implies, we already have our innate — or in-born — immune response at birth. It includes defensive structures such as the protective acid layer of our skin. This is effective because most pathogens are unable to survive on an acidic surface. Another example is the antibacterial enzyme called lysozyme, which is found in saliva and which attacks uninvited invaders before they can get any further than the oral cavity. The slimy surface of our mucous membranes, which contains bacteria-repellent substances, is also among those of our body's effective defence tools that form part of the innate immune system. After all, who can climb up a wall covered in slime without the aid of special equipment?

Most importantly, however, our innate immune system relies on specialised cells to keep us safe from dangerous pathogens. This army of cells protects us day in and day out. And like a real army, it's made up of specialist units and fighters, each of which is skilled in a certain form of defence — like the paratroopers, armoured corps, and marines of national military forces.

In a way, the front-line fighters are the granulocytes. Like other immune cells, they are a kind of white blood cell, or leucocyte. Unlike their red colleagues, they all have a cell nucleus, but don't contain oxygen-binding haemoglobin. Whenever they come across pathogens in the body while out on patrol, they immediately send out signals to their fellow granulocytes, which then deploy en masse and mercilessly massacre all intruders they encounter with the aid of special toxins.

The 'big eater' cells, or macrophages, make short work of anything that remains of the massacred pathogens. When infection occurs — that is, when disease-causing germs manage to infiltrate our organism — macrophages are attracted to the scene of the crime by regulatory proteins, where they gobble up anything that doesn't belong there. Should there be insufficient numbers for the mission at hand, they use a sophisticated messaging system to call in reinforcements quickly.

These mechanisms have developed over millions of years of evolution, constantly adapting to current needs and to the zeitgeist. For example, just like the FBI, the innate immune system uses a kind of profiler to identify dangerous intruders, only these individuals are known as 'natural killer cells'. A pretty cool name, too, I think you'll agree. Unlike FBI agents, however, these killers aren't particularly discreet. In fact, they behave more like Rambo than Clarice Starling. But these tough guys also specialise in identifying diseased cells and rendering them harmless, thereby nipping in the bud any tendency to develop into malignant tumours. They achieve this by ruthlessly forcing mutated or pathogen-infected cells to commit suicide — a process known biologically as apoptosis.

This is necessary because a certain proportion of the cells of our body are constantly mutating, i.e. altering the genetic information they contain. The next time they divide, the resulting daughter cells are sick, and if they were not recognised and rendered harmless as quickly as possible by the killer cells, they would eventually spell the end for us.

There's another, quite different reason why our cells need the ability to commit suicide. During our development as embryos in the womb, our fingers begin by being webbed, that is joined together by tissue. If this tissue didn't later disappear by means of targeted cell suicide, we would all come out bearing a certain resemblance to Ariel, the Little Mermaid. And this would have certainly made it more difficult for our forest-dwelling ancestors to swing between the treetops.

All these 'protectors' are part of the solid (that is, cellular) immune system. However, our body's defence system also includes fluid elements. Substances called plasma proteins circulate in great numbers in the plasma of our blood, on the lookout for unwanted intruders. In contrast to the cells described above, they don't rush headlong at a pathogen, but rather sneak up on it surreptitiously. Around 30 of these proteins make up what medics call the complement system. They attach themselves to microorganisms, enter inside them, and render them harmless. They're also able to expand blood vessels and call immune cells to their aid.

Our army of immune cells is constantly taking on new recruits. It finds them with the help of substances called interleukins. Among other things, interleukins promote the growth and development of white blood cells, as well as helping them divide and become activated.

All this might sound like an invincible army, but that is deceptive. Although the innate immune system is able to react rapidly to neutralise invaders, it isn't very innovative in the methods it uses. It always reacts in the same way, no matter whether it encounters a given pathogen for the first time or the hundredth. As long as its defence methods continue to be effective, that's no problem. But when that is no longer the case, the innate immune system is in urgent need of reinforcements.

The adaptive immune system

Which brings us neatly to the acquired, or adaptive, immune system. It is far more ingenious than the innate immune response, and is therefore able to react in a much more varied way. Not only because it's far more able to adapt to changing circumstances, but also because it's capable of learning. Every intruder carries particular markers on its surface, and the cells of the adaptive immune system are able to recognise these characteristic features. And, as if it weren't enough that they slay their opponents pitilessly, they also carve pieces off and display them like war trophies. This gives other immune cells the opportunity to commit their characteristic features to memory for a few years.

Thus, once a pathogen has come into contact with the cells of the adaptive immune system even just a single time, the immune system will reliably remember it from then on. It achieves this by creating so-called memory cells, which react at lightning speed as soon as they encounter the same intruder again. It does this by means of cells called B and T lymphocytes. The job of the B lymphocytes is to defend the body by targeting specific pathogens and other foreign substances. They do this by producing antibodies that immediately react to the characteristic markers of a familiar germ — remember those carved-off parts — then attach themselves to those pathogens and render them harmless. They could be described as the handcuffs of our immune arsenal.

You would like a bit more detail? Very well, here you are. When inactive B lymphocytes encounter a foreign substance (known as an antigen) as they circulate in our bloodstream, they waste no time in absorbing it, pulling it apart, and presenting the resulting fragments on their surface. This is a signal to the other type of lymphocytes, T helper cells, to begin releasing regulatory proteins. The 'T' in their name, incidentally comes from the name of the organ where these cells mature, the thymus, which is located behind the breastbone. It is particularly important during childhood and puberty. Later in life, it is less important in the maturation of T cells, and slowly shrinks until little more is left than a tiny island of almost inactive fatty tissue.

The proteins produced by the T cells activate the B lymphocytes, which quickly migrate to the lymph nodes and the spleen, where they start dividing like crazy. While doing so, they continue to manufacture many different antibodies until one happens to be produced that is perfect for combatting the pathogen in question. 'The more, the better' is the motto here. A tiny proportion of these B lymphocytes develop into the previously mentioned B memory cells.

In the final stage of their maturation process, the B lymphocytes eventually become plasma cells. They no longer have much desire to divide and so now produce only the antibody that was recognised as particularly fitting. Almost like a human being who doesn't mind continuing to work later in life but no longer has any desire to have children.

As I've shown, the immune system is pretty complex and sophisticated. However, it needs to be this complicated if it is to be able to protect our body effectively against intruders without turning on itself. It must be able to distinguish perfectly between its own and foreign cells in order to recognise pathogens as such straight away. A simple cut to the finger is enough to open the way for disease-causing germs to enter our bodily autobahn. If our body's army were not so skilled at halting the pathogens that are a constant threat to our blood vessels, our heart, and other tissues, and rendering them harmless, we wouldn't stay alive for very long.

One way we can support our immune system in its crucial mission is with immunisation. This makes use of the immune system's memory by presenting it with injected dead or weakened pathogens. They are harmless for us, but they still bear the same markers on their surfaces and so trigger the same complex responses from the immune system. Then if the pathogen in question should enter our body at some time in the future, our immune system is optimally equipped to deal with it. The memory cells easily recognise the troublemaker, and the plasma cells immediately begin producing masses of those antibodies that have been waiting for just this invasion since they were primed by the first infection. The intruders don't have a chance against the onslaught, and the immunised person remains healthy.

Just a Little Prick

'This vaccination terrorism has to stop! I'm certainly not letting anyone inject *MY* child with that poison.'

I'm sitting on a train to Berlin, eavesdropping on a conversation between two of my fellow passengers. Apparently, a young mother and her companion of about the same age.

'Yes, we didn't have Paul vaccinated,' the mother continues, 'and he's perfectly healthy, *AND* we didn't help to line the pockets of those money-grabbing multinationals.'

The young mother's friend nods in agreement.

'And the globules have always helped when he's needed them.'*

Of course, I can understand a mother's concern for her children's health, but anyone who manages to cram so many short-sighted old cliches into a single sentence has clearly not engaged seriously with the topic at all. A train carriage isn't really the place to intervene, so I bite my tongue, difficult as it is. But even among my own friends and acquaintances, the question of whether children should be vaccinated crops up all the time. Personally, I'm a great fan of vaccination, or immunisation as it's also called, because it protects us not only from diseases like polio, tick-borne encephalitis (TBE), and influenza, but also from myocarditis, or inflammation of the heart muscle.

---

* Meaning homeopathic globules.

'However,' the mother on the train might say, 'we should remember that not all vaccinations are the same.' It's true. There are not just different kinds of vaccination, there is also a broad range of different vaccines. But in principle, all vaccinations work by preparing the body for contact with pathogens in the future. There are two basic types: passive and active immunisation.

Active immunisation involves injecting people with dead or weakened versions of the pathogen they are to be protected against. The person's body reacts to them in the same way it would to living invaders. The immune system goes into overdrive, producing antibodies and memory cells, and commits to memory everything it needs to know to tackle a second infection. Although, it can take a few weeks before this reaction results in complete protection from the pathogen.

Passive immunisation is used when contact with a particular pathogen has already taken place, or may take place in the near future. The immune system doesn't have time to produce the necessary antibodies and so they are transferred to the person being immunised in a pre-synthesised form. After all, it wouldn't make much sense to inject people with copies of a pathogen that they already have inside their bodies. Thus, with passive immunisation, the antibodies that provide protection are not produced by our own immune system, but in the bodies of animals, such as chickens, pigs, horses, cows, or rabbits (which have been actively immunised). The advantage of this method is that the effect is immediate. But what are the risks associated with these procedures?

In both types of vaccination, the donor animals of the immunisation serum can pose a certain, if rare, problem. When the first active immunisation was developed almost 200 years ago against smallpox, hens' eggs were used to culture the antibodies — although no one yet knew the scientific explanation for the protective effect of immunisation, and the term 'antibody' hadn't yet been invented. In fact, hens' eggs still play an important part in the production of vaccines today.

Fertilised eggs are injected with living versions of the pathogen and left to develop for a while. The longer the chick — or rather the chicken embryo — is left to grow, the more pathogens are found in the egg. Eventually, the egg is opened up, and the pathogens inside are killed using chemicals, so that they are no longer capable of causing disease. However, fragments of them remain, and these are sufficient to prompt our immune system to produce the specific antibodies it will need to fend off future infection with that pathogen. But a vaccinated person can develop an allergic reaction to the residual hen proteins the vaccines contain.

Vaccines can also be produced in large bioreactors, where infected cells are cultured by planting genes from the pathogen into other microorganisms (such as bacteria or yeast). Once inside the host cells, those genes produce fragmented versions of the pathogen, which can be used for immunisation.

Even when a new vaccine has been proven to be effective in animal and human trials, and is launched onto the market, the quality-control process continues. In Germany, any complications in patients vaccinated by their doctor must be communicated immediately to the Paul Ehrlich Institute.* This organisation then decides whether any action needs to be taken. In 2001, a vaccine was completely withdrawn from the market due to frequent complications.†

One common way of avoiding the kinds of allergic reactions described above is to use monoclonal antibodies as vaccines. They are produced in the laboratory by means of a complex process. They bind much more specifically to the desired antigen than the antibodies from our B lymphocytes. Such a monoclonal antibody fits like a key into only one lock.

If a person feels unwell or feverish after being vaccinated, a distinction must be made between a normal reaction to the vaccine and a real complication. Even dead pathogens can make our body think it is sick. That manifests itself in slight feelings of illness or fever. When they experience this, some people tend to assume they have reacted badly to the vaccination and have developed serious complications. However, those mild side effects usually go away again as quickly as they came.

Potentially life-threatening diseases such as smallpox, the mere mention of which used to strike terror into the hearts

---

* This is the German government's federal institute for vaccines and biomedicines. In the US, this task falls under the remit of the CDC and the FDA; the UK counterpart is the MHRA — the Medicines and Healthcare-products Regulatory Agency; in Australia, it's the Therapeutic Goods Administration (TGA).

† A vaccine against tick-borne encephalitis (TBE).

of people in the past, have now been wiped out thanks to widespread vaccination drives. The number of new polio infections in Europe stood at just under 1000 in 1988, but by 2004 that number had been reduced to zero. The number of children contracting diphtheria and measles has fallen by 90 per cent.

We have comprehensive vaccination programs to thank for these developments in our parts of the world. And if an unvaccinated child doesn't contract diphtheria, this is due to the fact that there are now very few pathogens around. Luckily, other children's parents, who have not been blindsided by misleading propaganda, and prefer to trust the large body of scientific research showing the positive effects of comprehensive vaccination, continue to have their children immunised.

Vaccination is more or less essential for people with pre-existing conditions like diabetes or cardiovascular disease. Such people's immune systems are already less effective than that of a healthy person, and a relatively slight infection can trigger serious complications such as myocarditis in these people. Luckily, modern medicine has come up with targeted vaccines to protect against pathogens that can cause the heart muscle to become inflamed. But even a vaccine against genuine flu (influenza), tetanus, or diphtheria, for example, can be a real blessing for a diseased heart. Vaccinations not only protect our heart and other organs from dangerous infectious diseases, but also make the world a safer place for everyone. With just a little prick.

A Red Card for the Heart

Beep, beep, beep, beep, beep. It's 6.30 in the morning and your alarm clock is going off. You sit up in bed, but ... bwah! You feel like death warmed up. It looks like you've caught a cold. Yesterday you had a slight tickling in your nose and now you feel like a boxer's punching bag at the end of a long training session. And you haven't even tried getting out of bed yet. You struggle to your feet; your every move is painful. Maybe a painkiller will help? So you swallow a tablet and take a shower. Then off you go to work. It seems like the pill has done its job and you feel a little better. A good thing, too, because you have a desk full of work to get through at the office.

Which of us hasn't been there? Who has not taken an illness into work with them? With the noble aim of not disappointing the boss and leaving our workmates in the lurch. But is it healthy? Of course it isn't. And you're doing harm not only to yourself, but also to your co-workers, who you might find sniffling just like you at the office tomorrow.

It can't be as bad as all that, you think. Whether you let such a trivial cold keep you in bed or you pop a painkiller and drag yourself into work doesn't make all that much difference in the grand scheme of things, surely? You could hardly be more wrong! With your overactive sense of duty, you might just be setting the foundations for an inflammation of the heart muscle, or myocarditis. And it is anything but trivial. The pathogens that cause myocarditis do not attack the muscles of the heart alone, but the coronary arteries, too. This can weaken the entire organ to such an extent that it causes permanent heart failure, with all the unpleasant side effects this entails.

A serious case of myocarditis can even be fatal. It is extremely difficult to recognise and so hard figures about the frequency of deaths caused by myocarditis are lacking. Figures from Germany's Federal Statistical Office show that 3797 inpatients were diagnosed with acute myocarditis in 2012. The number of undiagnosed cases is likely to be far higher.

Myocarditis is so dangerous because it can strike anyone, irrespective of age. This is why we see cases again and again of apparently super-fit young footballers suddenly collapsing on the pitch in the middle of a game. Diagnosis: sudden cardiac death. It can be caused by a lingering flu infection, which is actually a harmless viral infection, but failure to rest sufficiently and recover fully can allow the virus to spread through the body and attack the heart. When this happens, any sporting activity places an extra burden on the heart, which may turn out to be the last straw.

If a patient gets enough rest, and recovers fully from the infection, myocarditis is very unlikely. In fact, there's an effective preventive measure. One good way to reduce the risk of developing myocarditis is to follow all your primary vaccinations as a child with regular boosters as an adult.

Those who also eat a healthy diet (as described in the chapter 'Eating to Your Heart's Content'), get enough sleep and do regular physical exercise, will both improve the performance level of their entire body, and protect it from all kinds of illnesses. Not least of all myocarditis.

Rhythmic Gymnastics for the Heart

The connection between sport, our hardworking blood cells, and a strong heart

Jump, Heart, Jump

Despite the regular appearance of headlines about professional athletes suffering sudden cardiac death, you would be hard pressed to find a medical professional who doesn't believe that physical exercise is good for the heart. Indeed, there is a unanimous opinion that physical fitness plays an important part in cardiac health. A large number of scientific studies agree that regular exercise lowers the risk of dying early from cardiac or vascular disease. It also makes us better able to cope with stress, and that is also good for the heart. But what kind of sport is best for us? After all, we want to do our bodies a favour, not do lasting damage to our joints or other parts of our body.

It's important that the sport you choose is varied and not a chore. Indeed, it should be fun. You can read all about 'bedroom sport' in the previous chapter of this book, and I highly recommended it as a form of physical exercise. But it is up to every individual to decide what is best for him or her.

For many, the ideal sport is running. Recently, a friend told me about the 'runner's high'. This is a kind of euphoric feeling often experienced by endurance athletes, especially long-distance runners. It's caused by happiness hormones (endorphins), which can suddenly make an athlete feel light and able to carry on running forever without ever getting tired. As someone who isn't particularly sporty, I decided I wanted to experience that, if only just once. There was only one thing for it: to try out an experiment on myself!

4.00 p.m.: Fired up with enthusiasm, I head up to the loft in search of my old jogging shorts and running shoes. After clearing a load of junk out of the way, I find them in an old cardboard box. *These shoes look almost new,* I think to myself, as I blow a layer of dust off them.

4.05 p.m.: I find a spider in one of the shoes. With my toes. While putting them on. I finally get over my disgust and solve the problem using the vacuum cleaner. Such trifling obstacles are not enough to stop me.

4.11 p.m.: I'm standing outside the house in my running gear, mentally preparing for the upcoming event. One of my neighbours appears and stops for a 'quick chat'.

4.55 p.m.: All necessary neighbourly information having been exchanged, my quest for the legendary runner's high can now begin. With admittedly mixed feelings, I set off towards the local woods.

4.57 p.m.: I begin to feel the first sensations of effort in my body. A tightening in my leg muscles especially. This is perfectly normal, I suppose. After all, I haven't been out running for quite a while. I will certainly not let it stop me.

5.01 p.m.: I'm now feeling muscles I didn't even know I had. It's not a pleasant feeling. But I'll wait and see — maybe the feeling will go away again.

5.04 p.m.: I feel the worst case of aching muscles in the history of the world coming on.

5.07 p.m.: I'm starting to come to terms with the idea that I will be bedridden for the next few weeks, at least, perhaps even for the rest of my life.

5.10 p.m.: The pain is now almost more than I can bear. I consider whether it might be less painful to stumble into a pothole deliberately and just destroy all my body's joints and muscles in one fell swoop.

5.11 p.m.: I am now on the lookout for suitable potholes, but instead I spot a bench. Break time! As I flop down on the bench, I suddenly remember my friend advising me to do a couple of push-ups during running breaks, to keep my momentum going.

5.12 p.m.: I lay face-down in the dirt of the forest floor, panting. Suddenly hearing voices, I push myself up with a groan, and start to count loudly enough for people to hear: '... 313, 314, 315 ... !' Seconds later, when the ramblers are out of earshot, I slump to the ground like a sack of potatoes.

5.15 p.m.: I am on my way home. Walking. No, limping.

If only someone had told me earlier that not everyone experiences a runner's high — and those who do are mostly highly trained athletes. And still it's only likely to occur when a practised runner pushes his or her body to the limits of endurance.

I console myself with the thought that I don't want to set any new Olympic records, just exercise my heart and circulatory system. Even without experiencing a runner's high, if needs be. I know that the heart reacts like any other muscle to regular training. It grows in size and strength. This means it can pump more blood to supply our muscles with the oxygen they are screaming out for during extended periods of exertion. And a well-trained heart not only performs better during exercise, but also pumps less often to supply the body optimally when at rest.

If you think of the heart as an engine, it becomes clear why (sudden cardiac deaths notwithstanding) athletes have a higher life expectancy than the less athletically inclined. An engine that's permanently running at the highest number of revs per minute will break down more quickly than one that constantly runs at a pleasant low speed. And the same is true of an unfit heart, which has to beat much more quickly to supply the body with blood than a trained one.

A sample calculation might be the best way to explain this. Let's say an untrained heart has to beat 80 times a minute on average, while a well-trained heart only has to beat 50 times in the period. After 70 years, the unfit heart will have beaten almost 3 billion times, compared to 1.8 billion beats for the fit organ. That's about 40 per cent fewer beats. Well, that sounds great. But is it as good as it sounds?

Don't people say sport is bad for you? After all, we often hear about the problems of sports professionals — especially at the end of their careers when they're not training as intensively as they used to — suffering from enlarged-heart syndrome and dying early. But this is only true, if at all, for top athletes. Experts are in no doubt that, for amateurs, sport is definitely not bad for you. On the contrary, it plays a major role in maintaining a strong and healthy heart. And should any of you still be worried about 'athletic-heart syndrome', the best advice is not to abruptly abandon sporting exercise after years of training, but to 'train down'; that is, slowly but surely reduce your training schedule. Then nothing untoward can happen.

A group of researchers in Manchester investigated the effect of sport on the pacemaker cells of the heart in rats. They had one group of rodents ('sports rats', if you like) spend an hour running on a treadmill every day for 12 weeks, while a second group (we might call them 'cage-potato rats') were allowed to avoid any kind of physical exertion. At the end of the experiment, the physically active sports rats had a significantly lower resting pulse rate than their lazy colleagues. The researchers found out that this was due to changes to the rats' sinoatrial node, the primary pacemaker of the heart, where ion currents through certain membrane channels enable the pacemaker cells to stimulate themselves. When the scientists investigated the genetic code of these cells, they found that the sports rats had far fewer genes for these ion channels, known rather amusingly as 'funny channels', than those of the lazy animals. From this they concluded that regular exercise had caused a permanent change in the internal structure of the heart's primary pacemaker.

So regular exercise makes our hearts bigger, stronger, more efficient, and slower. All-round better. Nevertheless, my 'runner's high' experiment failed to convince me. Sport might be good for the health, but my experience robbed me of any motivation to go out running again. This despite the fact that I used to be a good runner. In Year Five at school, anyway. Whenever I'd opened my big mouth too much and annoyed the bigger boys, I made it home in minutes!

The Fight-or-flight Rocket-propulsion System

What turned me into such a fast sprinter on the way home from school wasn't just my legs, but also a part of my nervous system. The autonomic part, to be precise, which can be thought of as 'the nervous system of the organs'. Although the sinoatrial node is the primary pacemaker in the heart, higher-ranking centres can considerably influence its activity via the autonomic nervous system.

The autonomic nervous system has two branches, with contrary effects: the sympathetic and the parasympathetic nervous systems. Together, they control most of our bodily functions, including, importantly, those of the heart. And although they're complete opposites, they complement each other perfectly. When we find ourselves in an emergency situation, our sympathetic nervous system immediately puts us on the alert. It dilates our pupils to improve our vision in bad light, it increases the activity of our muscles so that we're ready for a fight or so that we can run away more quickly, and it dilates our bronchial tubes so that we can breathe more efficiently. The American psychologist Walter Cannon coined the term 'fight or flight response' to describe this entire complex of reactions. What a perfect description! When I was running away from the big boys after school, I was in flight (that is, fleeing), and my sympathetic nervous system was my inexhaustible rocket-propulsion system.

The parasympathetic nervous system has the opposite, 'relaxing' effect, which is activated, for instance, after a large meal, when we fall into what some people call a food coma. It's that feeling of postprandial somnolence, to use the more scientific term, that causes us to flop onto the sofa in a stupor after a big meal. This happens when digestion has top priority for the nervous system. The parasympathetic nervous system winds down our body's overall level of activity and increases the supply of blood to the stomach, gut, and liver. This phenomenon has been dubbed 'rest and digest', along the lines of Cannon's famous coinage.

The sympathetic nervous system has several different effects on the heart. One is to accelerate the heart rate, and this can be directly ascribed to the influence of the sinoatrial node. It can also increase the force of the heart's contraction. The mechanism controlling this is the activation of something called beta-1 adrenergic receptors in the cell membranes. It also reduces the duration of contractions, so that the heart can beat more quickly — and I could run away from the big boys faster.

Luckily, the highly complex autonomic nervous system can be influenced with medical drugs, which is extremely useful in the treatment of chronic cardiovascular disease, and especially in emergency medicine. One well-known set of drugs of this type is the beta blockers, which, as the name implies, block the beta receptors mentioned above, thus lowering blood pressure and reducing the pulse rate. Another set of drugs that influence the autonomic nervous system are those extracted from plants of the digitalis family, notably the foxglove. These are used to treat patients with severe cardiac insufficiency because they increase cardiac contractility (the strength of the heart's contractions) while simultaneously reducing heart rate.

The dramatic case of a cardiac arrest calls for even more drastic intervention: medics increase the activity of the patient's sympathetic nervous system and simultaneously decrease that of the parasympathetic system by administering adrenaline and atropine. Adrenaline is known as a sympathomimetic drug; this term is a fitting one, given its function, since it comes from the Greek word *mimesis*, meaning 'imitate'. Adrenaline imitates the sympathetic nervous system, meaning it increases its activity, thus raising the patient's heart rate, expanding the bronchial tubes and increasing blood pressure.

Atropine, by contrast, acts as a parasympatholytic, which means something like 'blocker of parasympathetic effects'. Thus, it reduces the influence of the parasympathetic nerves on the heart. After these two drugs with similar effects have been administered, resuscitation efforts to restart the heart are much more likely to succeed.

The effects of all these drugs are by no means restricted to the cardiovascular system. In small doses, they can be useful in normal, everyday life. For example, some nose sprays contain adrenaline (a.k.a. epinephrine), which causes the blood vessels in the mucous membrane inside the nose to contract, rapidly reducing swelling. However, this effect only keeps working if the spray isn't used too often — otherwise, a so-called rebound effect kicks in, increasing the blood supply to the mucous membranes and allowing them to swell up again. No wonder so many people become almost addicted to such nose sprays.

Furthermore, atropine analogues* are used by ophthalmologists. When administered as eye drops, these inhibit the effects of the parasympathetic system on the eyes, which include the contraction of the pupils. As a result, the influence of the sympathetic system gains the upper hand and the patient's pupils dilate. This makes it easier for the ophthalmologist to examine the interior of the eye, especially the retina. An unpleasant, but completely harmless side effect for the patient is extremely blurred vision for a couple of hours.

Dilated pupils, especially women's, used to be considered an attractive feature, so many ladies in the past would use atropine-based eye drops. Atropine was extracted from a poisonous plant, the deadly nightshade, which explains this plant's scientific name, belladonna — 'beautiful lady' in Italian.

---

* In chemistry and pharmacology, analogues are chemicals that have similar functions or structures.

Seeing Red

What would an engine be without petrol? Nothing but a pile of useless metal. An engine is there to drive the car. And to do that, it needs a full tank of petrol. And our body's equivalent of the engine's petrol is our blood. Without the red stuff, nothing would work!

Because it has so many important functions, blood is quite often and quite reasonably called a liquid organ. On average, human adults have between five and six litres of it flowing through their veins and arteries. Blood is made up of both liquid and solid components. The liquid part, the plasma, makes up about 55 per cent of the blood in an adult male* and consists mainly of water, proteins, salts, and monosaccharides, but also a wide variety of many other substances. Another 44 per cent is made up of solid components, called haematocrit, which consists mainly of our various blood corpuscles and the specialist cells of the immune system.

A while ago, when I was playing with my little niece in her treehouse, she said something that really got me thinking. Clumsy as always, I had hurt my arm, and after she'd blown extensively on the graze to 'make it better', she said in surprise, 'Your blood looks a bit like tomato sauce.' How right she was becomes clear when you consider not

---

* This figure is slightly higher in women.

only the appearance of blood, but also its properties. It is red and viscous and, like sauce, contains sugar and other solid components. From a physicist's point of view, blood is a non-Newtonian fluid, which actually means nothing more than that it has different flowing properties than water. This is due to the fact that blood contains a lot of substances that, unlike salt in water, aren't dissolved in the plasma.

This kind of a mixture of a liquid and undissolved solids is called a suspension. In blood, this is seen in the fact that its properties change with its flow speed. The faster blood flows, the more emulsified the suspension becomes, which means the finer the mix of fluids (which are actually immiscible) becomes. This is due to the plasticity of our red blood cells. If you drop a spoonful of olive oil into a glass of water, the two liquids don't initially mix, and the oil forms a thick film floating on top of the water. If you stir it vigorously, the oil will be distributed throughout the water in the form of tiny droplets. What you then have is an emulsion. In the case of blood, when it's flowing fast, the corpuscles act almost like those droplets of olive oil in water.

The corpuscles in our blood include the previously mentioned red and white blood cells,[*] as well as cells known as platelets, which play a role in blood coagulation when we injure ourselves. If you are clumsy enough to graze your arm, they immediately begin a process to staunch the bleeding from the wound. They do this by quickly clumping together in great numbers and releasing a threadlike protein called

---

*   'The Body's (Almost) Invincible Army', p. 196.

fibrin. A fibrin strand is 1000 times finer than a human hair and is one of the most elastic substances known to biology. The fibrin strands form a dense net, which stops the wound from bleeding. Under certain circumstances, this mechanism can save lives.

Red Blood Cells in Yellow Jerseys

Why is Eritrea called Eritrea? Because it is on the Red Sea. The Greek word *erythros* means 'red'. That's why red blood cells are also called erythrocytes. The second part of the word comes from the Greek *kytos*, meaning a container or a jar. Numbering 24–30 billion, they're the most abundant cells in the blood of vertebrates. In humans, they lack a cell nucleus and are flat in appearance, with a dimple on each side, making them look a little like fruit drops. Their shape makes it easier for them to take up oxygen, since the distance between the interior of the cell and its outer membrane is much shorter than it is in a more spherical cell.

It's not a good sign if they begin to change shape. This can happen due to various toxins, vitamin deficiencies, or genetic defects, which force them to abandon their flat, round shape and become globular, cup-shaped, or even covered in spike-like projections, like a thorn apple. They also change their shape in the narrow capillaries, where the erythrocytes give up their cargo of oxygen and swap it for carbon dioxide; but this shape-shifting is deliberate and advantageous. This time, they stretch themselves out thinly so they can pass through the narrow vessels in single file.

As explained in earlier chapters of this book, the job of our red blood cells is to transport oxygen from the lungs to the tissues of the body and to carry carbon dioxide from there back to the lungs. The reason they're able to act like tiny little mules and transport oxygen is that they contain the red blood pigment haemoglobin, which makes up more than 90 per cent of their mass. This oxygen-binding protein has a compound containing a central iron atom, called haem, to thank for its ruddy hue.

But how does the haemoglobin know when it is time to offload its oxygen and take up carbon dioxide, and vice versa? Well, this is due to something called the Bohr effect, which ideally maintains the acid–alkali balance in our blood. The more carbon dioxide it contains, the more acidic our blood becomes, and vice versa. Thus, the blood in the vessels of the oxygen-rich environment of the lungs is more alkaline than that in the capillaries at the tips of our fingers, for example, where it contains more carbon dioxide. To restore the balance between these two gases and maintain as constant a pH* value as possible, the erythrocytes in our fingertips offload their oxygen cargo and replace it with carbon dioxide, while the opposite process of gas exchange takes place in the lungs.

Constantly carrying gases around the body is pretty hard work for our erythrocytes. This explains why they don't live long — after just four months they're broken down by phagocytes (or scavenger cells) in the liver, spleen, and bone marrow. That forces our body to produce new erythrocytes constantly to replace the dead ones. Our bone marrow produces about two million of them per second, amounting

---

* A scale to measure the acidity or alkalinity of a solution.

to a staggering 175–200 billion red blood cells streaming into our blood every day.

In adults, erythrocytes are produced in the red bone marrow, but in unborn babies it's the liver and the spleen that produce them. This process is controlled to a large extent by a hormone called erythropoietin, which has become widely known by its abbreviated form, EPO, because of the role it has played in Tour de France and other dubious sporting victories. As soon as certain sensors in the body register a lack of sufficient oxygen, the kidneys begin to produce more of the hormone. This in turn massively stimulates the production of red blood cells, with the result that the blood is able to transport more oxygen and the performance level of the body increases considerably. Competitive athletes like to make use of this effect by training for extended periods at high altitudes, where there's less oxygen in the air, thus forcing their bodies to produce more EPO and so increase the number of erythrocytes in their blood.

But there's a much easier way to make sure you're the first to cross that cycle-race finishing line: you can simply inject EPO directly into your bloodstream. Of course, that is doping, and, as such, strictly forbidden in sport. It is not only cheating and unfair, but also highly dangerous for the doped athletes. Artificially increasing the number of red blood cells has the effect of making the blood more viscous, considerably increasing the risk of heart attacks, strokes, and organ damage. A mixture of EPO and stimulants is a dangerous cocktail, and those who take it may end up paying a high price for that yellow jersey. Indeed, we repeatedly hear of doped elite athletes dying of heart attacks at an early age.

Basically, no amount of doping substances can replace long years of intensive training. The hearts of athletes who take human growth hormones are far more muscular than those of normal human beings, but that muscle growth is directed inwards, which eventually reduces the capacity of the heart's chambers. A much healthier way to promote the growth of the heart muscles more gradually is regular training. This makes the heart stronger and eventually able to do great things — without the aid of illicit drugs.

The Pressure Is On

The mechanisms of blood pressure

Checking the Boiler Pressure

The phrase 'blood pressure' is used to describe the pressure exerted by the blood on the walls of the vessels in a given part of the vascular system. There are two different values to be aware of: systolic and diastolic pressure, often described as the upper and lower blood-pressure numbers. After the left ventricle has filled with blood, it contracts and pumps the blood into the aorta. This process is known medically as systole, and so the pressure in the blood vessels created by that process is known as systolic pressure. Another way of thinking of it is that the systolic pressure is the maximum pressure with which blood is pumped out of the heart to the rest of the body. Once this has happened, the left ventricle needs to fill with blood again. At this point, the pressure falls, of course, and the lowest pressure reached at that point is known as the diastolic blood pressure.

Both these values can be measured using a blood-pressure cuff and a stethoscope on the arm. If you place a stethoscope on the artery in the crook of your elbow, you'll initially hear nothing. This is because the blood is flowing through it unhindered. Flowing liquids only create a sound when they encounter resistance, a little bit like a brook that only begins to babble when its water splashes against some stones. Which is where the blood-pressure cuff comes in. Placed around the upper arm and firmly inflated, it creates

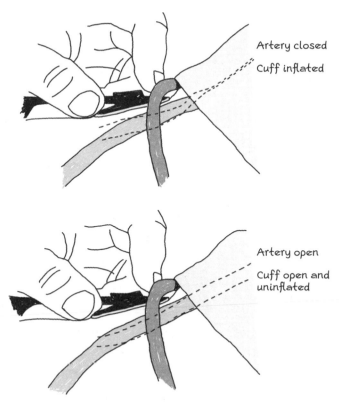

Artery closed
Cuff inflated

Artery open

Cuff open and
uninflated

Measuring blood pressure. The lines show the blood vessels in the arm.

an increasing amount of pressure on the arm and therefore also on the artery in the crook of the elbow, until this is eventually squeezed closed, completely obstructing the flow of blood. The valve of the cuff is then opened and it is slowly allowed to deflate — reducing the pressure exerted by the cuff. When that pressure exactly matches the pressure inside the artery, a surge of blood passes through the constriction before it closes again.

The quick rush of blood creates an audible sound. As soon as it's heard, the blood pressure is taken from the dial or display on the pressure gauge — this is, of course, the systolic pressure. A healthy value should be about 120 mmHg.*

When more air is now let out of the cuff, each surge of blood can be heard as a knocking sound as it forces open the still somewhat constricted artery. This continues until the blood is able to flow again unhindered. Then no more sound can be heard, which is when the diastolic pressure can be read. It should normally be around 70–80 mmHg. Thus, a blood-pressure reading of 125/80 (read aloud as: 125 over 80) means there is a maximum pressure of 125 mmHg in the arteries, and a minimum pressure of 80 mmHg.

A value measured at rest that's higher than 140/90 mmHg is considered mildly high blood pressure — or mild hypertension, to use the medical term. A value of 160/100 mmHg or above is considered grade 2 hypertension, and 180/110 and higher is grade 3, or severe hypertension.

A phenomenon that's often observed in doctors' practices and hospitals is called white-coat hypertension. Patients' blood pressure often increases enormously when they enter a hospital or doctor's surgery, although it may be normal under other circumstances. This effect can be excluded by means of a long-term observation over several days using a mobile automatic blood-pressure gauge.

Permanent high blood pressure increases the risk of

---

* This abbreviation stands for millimetres of mercury. Formerly, blood-pressure gauges used actual columns of mercury, like those in old thermometers. The value was taken by measuring how high, in millimetres, the mercury was pushed up the column by the pressure of the blood.

damage to the vascular system. One danger is that an artery in the brain may be stretched to such an extent that it tears, or even bursts like an overinflated balloon. This will cause blood to spurt out, the amount depending on the size of the vessel that bursts, and this can have serious consequences, even as far as sudden death. Blood vessels that are constantly exposed to excessive pressure are permanently stressed and may eventually suffer irreparable damage.

In emergency medicine, it's often said that the systolic pressure should be above 100 mmHg and the diastolic pressure below 100 mmHg, and if this is the case, the patient is doing well. That's true for most people, but it shouldn't be seen as a blanket rule for all questions relating to blood pressure. For example, if someone has chronic hypertension, let's say with a value of 180/110, and they suffer a circulatory shock, then their blood pressure will fall rapidly, for instance to a value of 130/70. Although the shock lowers their dangerously high blood pressure, this is anything but a blessing for the patient; on the contrary, it can be highly dangerous. In this context, 'shock' doesn't mean a state of high agitation or nervous tension, but a discrepancy between the volume of blood the body requires and the amount available — hence the medical term, hypovolaemia — leading to an insufficient blood supply to the capillaries.

This means the body's tissues are no longer properly supplied with blood, which can lead to immediate death if it affects the brain. Medical experts differentiate between absolute and relative hypovolaemia. Absolute hypovolaemia refers to a decrease in the overall volume of blood in the body, due to heavy bleeding as a result of a serious accident, for example. With relative hypovolaemia, the overall amount of blood doesn't change, but much of it sinks to the legs and other tissues of the lower body. Initially, the result of both kinds of hypovolaemia is the same: the body's organs receive an insufficient supply of nutrients and, most importantly, oxygen.

One common rapid way of helping in cases of relative hypovolaemia is to raise the patient's legs, to promote blood flow back towards the heart. Even in cases of absolute hypovolaemia, such a manoeuvre might help, but it's much more important to stop the patient bleeding as quickly as possible. Anyway, the next time you feel like putting your feet up for a while, you have a great excuse: you can say that your circulatory system is playing havoc with you, then lie back and enjoy a well-earned rest.

Bottles on the Lawn

It's a sunny summer afternoon and I'm sitting outside the ambulance station, enjoying the feeling of the warm sun on my skin. It has not been a busy day. A couple of patients to transport, two harmless call-outs, no complicated cases. Just three more hours till my shift ends. I lazily watch the twittering birds, and listen to the buzzing and humming in the bushes. But then my summer reverie is rudely interrupted as I feel a vibration on my belt. 'Not now ...' I beg, while jumping to attention. Luckily, it is not my pager, but my mobile phone. It's a text message from a friend inviting me to a barbecue that evening at her house. I slump back down on the bench and squint into the sunshine in relief. Just as I'm replying, the pager on my belt starts to go off after all. An insect bite with circulatory dysregulation, probably an allergic reaction. The patient urgently needs an ambulance, i.e. us, and an emergency doctor, who will probably have to come from the next village.

Stefan and I jump into the ambulance and within seconds we're speeding to the call-out, blue light on and siren wailing. Four minutes later, we arrive at the address and see a man standing in the road, waving. I get out of the ambulance, shouldering my emergency backpack as usual, and grab the oxygen-therapy bag in my left hand and the ECG in my right. We follow the agitated man round the back of the house into the garden. There, we discover an elderly lady lying on the lawn next to a flower bed. She's conscious; her breathing is a little more rapid than normal, but far from dangerous. What's more striking is her skin, which is as white as a sheet. When she sees us approaching, she points to her hand, whispering, 'Here! Stung here!' As I hand the woman an instant cold compress, Stefan quickly ascertains her medical history while he checks her blood pressure. 'Pulse is steady, barely palpable, tachycardic,' he sums up briefly while I prepare an infusion and search the woman's arm for a suitable vein.

'Steady pulse' is good news. The fact that it's difficult to feel in her wrist is to be expected in her state. Tachycardic means her pulse rate is high.

'Pressure 120 over 80.' My ears prick up. 120 over 80 sounds like a normal reading, but, seeing how pale she is, this woman's blood pressure can't be normal; it should be much lower than her normal value. It's at this moment that Stefan asks her about any previous conditions. Guiltily, the lady tells us she suffers from high blood pressure, but didn't take her medication that morning. She tells us her normal resting blood pressure reading is 190 over 110. Now it all makes sense.

Blood pressure too low, pulse rate far too high, together with an insect sting — everything points towards one diagnosis: anaphylactic, or allergic, shock. When this happens, the patient's blood vessels expand as a reaction to the toxins in the insect's sting, the walls of the blood vessels become porous, and fluid seeps out into the surrounding tissue. The most visible sign of anaphylaxis is weal-like swellings on the skin. And this lady has them in abundance.

When a shock like this occurs, the patient's organs are starved of blood. Especially if the patient is standing upright, too little blood makes it as far as the head; and when the brain is undersupplied with blood, the patient faints. This may be unpleasant for the patient, but from the brain's point of view it's a clever trick to get the body horizontal so that more blood can flow to the brain again. It is easy to understand this effect if you compare the patient to a bottle of water. The lid of the bottle is the patient's head, the middle of the bottle is the abdomen, and the water in the bottle is the patient's blood. If the bottle is filled to the brim, the water reaches every part of its interior. If, like a body in shock when not enough blood is available, the bottle is only half full, the lid will remain dry. If you tilt the bottle to one side, a little water at least will reach the neck of the bottle. Transfer that to a human patient, and it means a lying patient is hardly likely to lose consciousness, while a standing patient may faint. As mentioned, one way to improve the blood supply to the head and the brain is to raise the patient's legs.

This is, of course, what we did with our patient, and, once the infusion was flowing, the fluid helped replace the missing blood, and the colour began to return to her cheeks. Then we heard the siren of the emergency doctor's car. Luckily, everything turned out fine, and we didn't even really need the doctor's help.

Apart from (mostly involuntary) lying down, the body has other tricks to prevent blood pressure from falling too far. One less conspicuous system for regulating blood pressure is the renin-angiotensin-aldosterone system (RAAS), which helps contract the blood vessels and increase the volume of blood. It follows logically that when the heart beats more forcefully and pumps more blood into the blood vessels, blood pressure rises. The diameter of the blood vessels also influences blood pressure, of course; the narrower a blood vessel becomes, the more resistance the flow of blood will encounter and the higher the blood pressure will be.

Angiotensin II is the central hormone of the renin-angiotensin-aldosterone system. Before it's released, there is already a hormone precursor swimming around in our blood plasma: angiotensinogen, which is produced in the liver. Another hormone-like enzyme found in blood plasma is renin. This can be produced by the kidneys, the adrenal glands, the uterus, the saliva glands, or the pituitary gland. When the two meet, renin breaks down angiotensinogen to form angiotensin I. This is then converted into an active form (creatively dubbed angiotensin II) by an enzyme called ACE.* One of the effects of this angiotensin II is to make the smooth muscle fibres in the blood-vessel walls contract, increasing resistance in the blood vessel. It is the blood vessels' drill instructor, shouting at their walls: 'One more push-up, come on! I want to see you sweat!'

Angiotensin II also makes us feel thirsty, and stimulates

---

* Angiotensin-converting enzyme.

our appetite for salty foods. It causes the adrenal cortex to produce a steroid hormone called aldosterone, which prompts the kidneys to retain sodium and chloride ions. The result of this is that more water remains in the body rather than disappearing down the toilet. All these individual effects combine to increase the volume of blood in the body, and therefore also increase blood pressure.

As mentioned, the enzyme ACE plays a pivotal role in this complex mechanism to increase blood pressure. Contrariwise, blood pressure can be lowered by drugs that block this enzyme. This is the principle used by a range of medicines known as ACE inhibitors.

One of my favourite hormones, in the release of which angiotensin II is also involved, is antidiuretic hormone (ADH), which comes from the pituitary gland. It causes more water to be retained in the kidneys rather than being expelled from the body as urine (hence the name). Once again, the effect of this is to raise blood pressure.

In practice, it works like this: you're sitting in the pub with a belly full of beer, keeping your legs tightly crossed. The unpleasant feeling in your nether regions has now developed from a slight twinge to a burning pressure. You know due to the results of previous experiments on yourself during evenings at the pub that you are doing the right thing by not going to the toilet straight away, because you'll only have to go again in about 15 minutes, anyway. But the pressure is starting to become excruciating. Soon, you can no longer stand it: you go and empty your tortured bladder with a sigh of relief. The reason beer and spirits have us heading to the toilet so often is because alcohol inhibits the aforementioned ADH. Then our kidneys retain less water in the body, our bladder fills up more quickly, and we turn into a human waterfall.

In the short term, the heart can react to increased blood pressure by using its right atrial appendage, which is a pouch-like extension in that chamber of the heart. When it is extended by an increase in the volume of blood, it releases a hormone that causes the kidneys to secrete sodium chloride along with water. The result is a drop in blood pressure.

However, when levels of angiotensin II are elevated for a longer period of time, the result is usually high blood pressure, possibly causing organ and blood-vessel damage. For this reason, hypertension should always be treated with drugs, such as ACE inhibitors, where possible. Another medicinal option is provided by beta blockers, which reduce the contractile power of the heart to lower the pulse rate and blood pressure. This should always be done with care, especially when the heart is already weakened. A weak heart should be treated with very small doses at first; the dosage can then be raised as long as the patient shows no ill effects. If this procedure isn't followed, there's a danger of lowering blood pressure too far.

We can influence many of the factors involved in high or low blood pressure, such as our diet, drinking, and smoking, by making lifestyle changes. Others cannot be influenced in this way at all. According to some research, there's an apparent link between hypertension and birth weight. Thus, babies with low birth weight are more likely to develop problems with high blood pressure later in life than heavier babies. It seems that lighter babies begin with conspicuously low blood pressure, which then rises more quickly during the first year of life than that of heavier newborns. The reason for this seems to lie in the phenomenon of catch-up growth. The faster a small body tries to make up for lacking growth, the greater the likelihood is that it will develop cardiovascular problems, which lead to blood pressure that's high enough to require treatment.

Pregnant women often have problems with fluctuating blood pressure. Although common, this is rarely dangerous. Pregnancy is a wonderful natural phenomenon, and women's hearts are actually well prepared for it — able to beat not only for the mother, but for her unborn baby as well.

Beating for Two

Pregnancy is an incredibly exciting time for the parents-to-be, and can mean a lot of heart palpitations for both. Will everything go well? What will our life as parents be like? Can you feel it kicking, too? It never fails to impress me whenever I see such a little miracle growing inside another human being. So much change in such a relatively short time. And by that I don't just mean my mate who had to swap his sports convertible for a 'sprog hopper' when he was about to become a father. He even painted his 'man cave' pink and turned it into the perfect nursery. Those were two sacrifices he was more than happy to make. His wife had to make much more serious sacrifices. One of the issues she had to deal with during her difficult pregnancy was problems with her blood pressure.

Around a quarter of expectant mothers develop pregnancy-related high blood pressure from around the 20th week.* It particularly affects pregnant women who are overweight. It isn't a dangerous condition, however, as long as it's permanently monitored by a doctor; although if the mother-to-be's blood pressure becomes too high, there's usually no alternative to hospitalisation. There, doctors will treat the condition with medication. There's no need to worry about those drugs harming the unborn child; after all,

---

* Doctors call it gestational hypertension.

the doctors' highest priority is to make sure both mother and child remain as healthy as possible.

In most cases, the mother's blood pressure returns to normal, pre-pregnancy levels within about three months after the birth of her baby. Scientists need to do more research to discover exactly why a woman's blood pressure should be so liable to rise during pregnancy. One theory is that it has to do with the increased volume of blood in her body. During pregnancy, of course, the mother's heart is beating for two.

But what about women who have a heart condition? Pregnancy places an immense burden on a woman's entire body, and this includes her heart. Not only because the volume of blood in her body increases by about half, but also because a woman's heart gets bigger when she's carrying a child. This is understandable, since it has to work a lot harder. As mentioned, fluctuations in blood pressure are common during pregnancy, but for those with a weakened heart they can become a serious problem. If the expectant mother's heart is overworked and can't pump enough blood around her body, her baby is in danger of being undersupplied. This can lead to the embryo being underdeveloped and being born prematurely, or even stillborn. To be on the safe side, some gynaecologists may even advise women with certain heart conditions against getting pregnant at all. These include women who have undergone heart-valve replacement* and

---

* An artificial heart valve is surgically implanted to replace a natural one when the valve no longer closes properly, or becomes too narrow.

those with Marfan syndrome, a disorder of the connective tissue with many negative effects on the heart.

One condition receiving increasing attention among medical researchers is called peripartum cardiomyopathy (PPCM). This complicated-sounding medical term simply means 'disease of the heart muscle around the time of giving birth'. It describes a phenomenon in which women with otherwise healthy hearts begin suffering symptoms like fatigue, difficulty breathing, unexplained coughing, swollen legs, and heart palpitations. This dramatic situation can culminate in a life-threatening cardiogenic shock.

Scientists currently have no idea what causes this condition, and it's the subject of intensive research at the Hanover Medical School. One theory is that PPCM is a disease of the blood-vessel walls, which could be caused, in part at least, by the milk-producing hormone prolactin.* Risk factors like high blood pressure, smoking, and infection also appear to play a part in PPCM.

Scientists in Hanover are currently investigating ways to treat PPCM with drugs, using a medication called bromocriptine, which inhibits the hormone prolactin. The scientist leading the study says, 'Although the criteria for the appearance of PPCM are clearly defined, this condition often goes unrecognised.' This may be due to the fact that during pregnancy, when their bodies are going through such huge changes, many women feel a bit unwell generally, and so

---

* Prolactin not only enables milk production, but also helps the uterus to return to its normal state after childbirth.

simply don't recognise the symptoms of PPCM as such.

Thankfully, the condition usually gives no more cause for concern once the baby is born. On the contrary: then everyone has come through the birth more or less unscathed, a new life has begun, and another little heart is now beating merrily and healthily away in its owner's body.

Sleeping Beauty's Heart

Of (un)healthy sleep, too much stress, lovesickness, and heart defects

The Heart Can't Sleep

I lie awake in bed, listening to the tick of the alarm clock. Why, oh why can't I get to sleep? Of course, I'm still pretty wound up — today was a stressful day. No, actually that was yesterday, because it's now already half past three in the morning. In less than three hours, that stupid thing will be going off again. I turn over onto my right side, then back onto my left. I then spend at least ten minutes trying to find the perfect position for my pillow. I just can't settle.

I was reading about the connection between sleeping problems and heart failure only a week ago. A group of Norwegian researchers published an article in the *European Heart Journal*, describing the results of an interesting study they carried out that involved following a total of 54,000 people aged between 20 and 90, over a period of 11 years. The study failed to prove conclusively that those who have trouble sleeping have a higher risk of heart failure, but it also left open the possibility that this may be the case. This is because insomnia is a source of stress, and stress triggers the release of a range of hormones, all of which have a negative effect on the heart. Over the long term, that could certainly be a contributing factor to chronic heart failure. Over 1400 subjects in the Norwegian study suffered from some kind of cardiac insufficiency, and it was striking how many of them also had difficulty getting to sleep or staying asleep through the night. But who is to say that their heart problems were not caused by other factors?

To exclude that possibility, the researchers examined every subject's life habits in detail, and recorded their blood pressure, cholesterol levels, physical activity, and tendency towards depression and anxiety. They also took body size and weight into consideration. All these variables were then factored out of the calculations. However, the subjects were never examined in a sleep laboratory to rule out other possible conditions that carry a risk for the cardiovascular system, such as sleep apnoea.* Despite this, they were able to conclude that serious sleeping problems may well have a negative influence on the heart.

To see how this might be the case, we need to understand the different sleep phases we pass through once our heads hit the pillow and our eyes fall shut.

The first phase is the onset of sleep. Falling asleep can take some time, depending on how tense we are and how active our body still is. During sleep onset, we slowly drift away from the waking world, but we can still be awakened easily and return to reality. If we're lucky enough to be left in peace, our heart rate begins to slow down, our blood pressure sinks, and our breathing becomes more regular. Our muscles and, importantly, our minds relax, and we are now perfectly prepared to enter the next phase of sleep.

The next phase lasts only a few minutes. Although our muscles continue to relax, we sometimes twitch violently

---

* Suspension of breathing during sleep.

once or twice during this stage. Anyone who shares a bed will certainly be familiar with this, when their bedfellow 'jumps' as if startled, and you may even have experienced a painful kick in the shins or two.

Despite this occasional twitching, our heart begins to beat even more slowly, and our blood pressure sinks further. Our eyes begin to roll slowly beneath our eyelids and we gradually slip into the third, deeper phase of sleep. In this stage, our eyes remain more or less still and we are deeply relaxed. However, it's possible in this phase that we will relive the mental conflicts of the previous day, which at the time may not even have seemed important. The more traumatic they were, the more often they are replayed, and our heart beats a little faster each time. Only when that is done can we pass on to the fourth phase of sleep.

In this phase, brain activity is reduced further and our sleep becomes deeper and deeper until we eventually reach phase five: absolute deep sleep.

This is the period of our nightly sleep when our body regenerates best. Our heart now beats only very slowly, as few as 50 times a minute for some people, our blood pressure is at rock bottom, and we are completely relaxed. At this point, our body is both resting and regenerating. Our immune system in particular takes advantage of this opportunity to regroup. Only in this way can it be ready to protect us optimally after we get up the next morning. This is why those who don't get enough sleep will always get sick more often than those who do.

The deep-sleep phase lasts between one-and-a-half and two hours, and is repeated several times through the night. If we're woken during one of these phases, we find it most difficult to get up, and are usually in a foul mood when we do. Given the opportunity to go back to bed, we usually fall asleep again straight away.

As the morning gradually approaches, our deep sleep is interrupted increasingly often, and we pass into the REM* phase, characterised by hectic eye movements, and a rise in brain activity, blood pressure, and pulse rate. This is when we dream and our bodies and minds process our experiences of the waking world. Without this mechanism, we would find it impossible to deal with stress, leading to serious psychological and physical problems.

Does this mean we have the perfect excuse to snooze in bed till late in the afternoon? Unfortunately, no. How fit would Sleeping Beauty's heart be? As every child knows, she was a world-champion sleeper. Was that good for her heart? Probably not. It's a wonder that she woke up looking as fresh as a daisy after sleeping for 100 years, because too much sleep is unhealthy. This is down less to the sleep itself than to the associated lack of physical exercise, which would have been extreme after the century-long slumber of our fairy-tale princess. Actually, she must have woken up with extremely damaged blood vessels and an inability to move any of her limbs. A body that never moves becomes increasingly weak over time. One hundred years of sleep would not keep you

---

*   REM stands for rapid eye movement.

young and fresh. Even if you were romantically awakened by the kiss of a handsome prince at the end of it.

Researchers at the University of West Virginia found that people who regularly sleep for more than nine hours a night have an almost 50 per cent higher risk of suffering a heart attack or other cardiovascular disease than those who are less slumberous. According to that study, the ideal amount of sleep from the point of view of cardiac health is seven hours. Sleeping fewer than five hours a night, however, can as much as double a person's risk of heart disease. So, both sleeping too much and sleeping too little can be harmful to our health. Seven is the perfect number!

The Lovesick Heart

Falling in love is one of the most beautiful and exciting feelings in the world. We need only to think of our beloved partner and our heart skips a beat. We are full of energy and plans for the future. And a particularly pleasing side effect is that the happiness and bonding hormones our bodies produce when we're in love are good for our long-term cardiac health. Scientists at the University of California have even found out that the heartbeats of couples in love become synchronised, even when they are simply sitting opposite one another, staring into each other's eyes. They also noticed that women's hearts appear to be quicker to synchronise with those of their lovers than men's. Exactly why this should be remains a mystery, but isn't it totally romantic to think that lovers' hearts really do beat as one?

Yet what about when the romance dies? What happens to our heart then? Anyone who's been properly lovesick knows how painful it is to feel so lovelorn, so sad, and so worthless. It seems to us that our loss is too great to bear — the pain is almost too much. Just getting up in the morning, having a shower, and facing the day takes a huge effort of self-will. We have no appetite. It feels as if the sun has sunk below the horizon forever.

We see ourselves in the leading role of a tragedy worthy of any stage, and we experience feelings of desperation, as if we are about to fall apart, not just emotionally, but physically, too. But can such emotional stress kill us? Is it possible to die of a broken heart?

The answer is: yes, it really is possible. Although it's very rare. Feelings of grief, protracted sadness after a loved one is gone, and long-term emotional stress can have a huge effect on the physical body. It doesn't even need to be a drastic, life-changing event to eventually cause us physical harm. Lack of appreciation, bullying, or constant complaining are enough to drive a person into such a gratification crisis,* which can have a massively detrimental effect on physical health. The source of the pain may be in the mind, but the pain itself is very real and by no means imaginary.

Back pain is an extremely widespread medical complaint in industrialised nations. Even just the fear of back pain makes us change the way we move, adopt protective postures, and tense up. And before we know it, we're in a vicious circle. Only one person in five in modern industrialised countries reports never having suffered from back pain; the remaining 80 per cent say they've had back pain at least once in the past. According to studies carried out around the world, back pain costs many billions of dollars, pounds, and euros every year. Back pain can be caused by heavy physical work, but also by emotional stress. Disaffection and stress in the workplace can be a huge factor in back pain and similar conditions, however too much emotional stress can influence not only

---

* A gratification crisis is caused by a lack of sufficient rewards after great expenditure of effort and sacrifice. It can lead to mental illness.

our bodily posture, but also our hormone balance and with it the functioning of our internal organs.

A condition that was first described relatively recently is interesting in this context: takotsubo cardiomyopathy, which is a stress-related alteration of the muscle of the heart. It's also known as broken-heart syndrome, transient left-ventricular apical-ballooning syndrome, and stress-induced cardiomyopathy. It occurs predominantly in post-menopausal women following exposure to sudden, unexpected emotional or physical stress. The dysfunction of the heart muscle caused by this syndrome is associated with symptoms — such as acute difficulty breathing and severe chest pain — that are very similar to those of a heart attack. An ECG will often show an ST-segment elevation, which is also a typical sign of a heart attack. Takotsubo cardiomyopathy is also associated with cardiogenic shock, rapid irregular heartbeat, and even ventricular fibrillation. All of these symptoms are immediately life-threatening and must be treated as quickly as possible.

In such cases, when doctors in the catheterisation laboratory examine whether there is a narrowing of the coronary arteries, it usually turns out that the problem doesn't stem from the coronary arteries at all. The problem is a deformation of the left ventricle of the heart. This chamber is no longer able to carry out its pumping duties — almost as if it were paralysed. When the patient's life circumstances improve, the condition usually goes away. As long as they receive immediate and intensive treatment, patients are

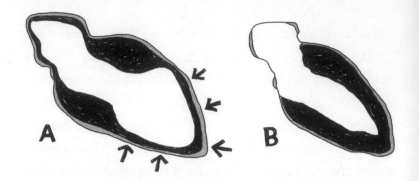

In takotsubo cardiomyopathy, or broken-heart syndrome,
(A) the muscles at the apex of the heart no longer contract properly.
(B) shows a healthy heart for comparison.

usually healthy and able to handle stress again within about a month. Only around 1 per cent of cases end fatally.

On 23 October 2004, Japan was struck by a severe earthquake, measuring 6.8 on the Richter scale. In the wake of this natural disaster, Japanese researchers closely followed 16 patients diagnosed with this syndrome. They called it takotsubo cardiomyopathy since the deformation of the ventricle reminded them of the shape of traditional takotsubo pots used for catching octopuses in Japan.

The study group was made up of 15 women and one man, with an average age of 71.5 years, all of whom had experienced the quake firsthand. The researchers calculated that the stress reaction triggered by the earthquake increased the likelihood of takotsubo cardiomyopathy by 24 times. However, it remains unclear why the vast majority of people affected should be women. One theory puts it down to women's generally more emotional state, but that seems like a rather rash conclusion to me.

Another attempt to explain this phenomenon is based on the fact that many patients in such situations have increased levels of stress hormones from the adrenal cortex, primarily adrenaline and noradrenaline, in their blood. It's suggested that the low levels of oestrogen in the blood of postmenopausal women make their hearts more susceptible to the effects of those hormones. But this also seems rather far-fetched to me.

Incidentally, more positive experiences, for example winning the lottery, can also trigger broken-heart syndrome. This would make an association with increased levels of adrenaline and noradrenaline more probable, since the effects of a certain kind of tumour in the adrenal glands, called a phaeochromocytoma, which also triggers the release of large amounts of those stress hormones, are similar to those of broken-heart syndrome.

Researchers have suggested that the concentration of a protein called sarcolipin in the muscle cells of the left ventricle could be significant. This protein inhibits the transport of calcium, which is important for the contractility of muscles. If too little calcium is present, the muscle contraction is significantly weakened.

Another group of researchers have discovered that adrenaline can have the reverse effect, increasing the contractility of the muscle cells by binding with beta receptors on their surface. When this happens, certain proteins increase the contractile force of the muscles via a chemical-reaction cascade inside the cells. When the researchers injected mice with very large amounts of adrenaline, those receptors changed, with the result that other proteins significantly reduced the contractility of the muscle cells — again via a complex reaction cascade.

Why should that be the case? One theory says it's because the body is trying to protect the heart from damaging influences, but simply overshoots the mark in broken-heart syndrome. And the reverse adrenaline effect would then be a mechanism to protect the muscle cells from overstimulation by stress hormones during an emotional emergency. After all, such an overstimulation can also have fatal consequences.

As you can see, there are many research projects and theories aimed at explaining the cause of broken-heart syndrome. In 2004, within the framework of a study of two sisters with broken-heart syndrome, it was suggested that a risk of developing the condition was genetically controlled. Literature from 2006 suggests the condition is triggered by infection with human herpesvirus 5. And so on. However, since the number of described cases is so small, they remain mere theories to this day. To tackle this problem, an international takotsubo registry (InterTAK) was set up in 2011, in which 26 centres around the world participate. The registry now contains data sets on 1500 patients.

Hopefully, such registries will one day enable researchers to find out more about the cause of this syndrome and help develop best-treatment policies. But no matter how successful the scientific research is, there's one thing it will never be able to do: ease the pain of the broken-hearted. When we are really lovesick, no doctor in the world can take away those terrible feelings. The only thing that might help a little is the support of good friends, who can offer a shoulder to cry on. And remembering that time is a great healer.

Wallowing in lovesick self-pity is often seen as a weakness, but this is an attitude I don't share. What could be greater proof of a person's humanity than to be moved to the core by feelings of love? Or the loss of it?

Tic Tac Heals All Wounds

Why do we experience lovesickness as being so terrible and torturous? Part of the reason is certainly because, in our desperation, we simply cannot imagine the pain will ever subside. This makes us feel it all the more keenly. Medical experts call this the 'nocebo' effect. That word is Latin and means 'I shall harm'.

The nocebo effect explains why the expectation of serious complications following a vaccination can lead some patients to perceive a minimal physical reaction as the worst illness of their entire life. This reaction is usually not deliberate on the patient's part, and it's certainly not all in the mind. The phenomenon has real, measurable physiological results. There's even a story of a man who suffered a fatal heart attack because a voodoo doll representing him was 'critically injured' and he believed so strongly in voodoo magic. This may just be a good story to tell down the pub, but there are many verifiable examples of the nocebo effect in the medical literature.

Doctors are familiar with the phenomenon that patients who have been given extensive information on the possible side effects of the medicine prescribed to them tend to be more likely to experience those side effects than patients who take the same medication but are not provided with that information. The most interesting aspect of this effect is that it doesn't seem to matter whether the medication they're given contains any active ingredients or is simply a dummy pill.

The same is true of the opposite effect. A couple of years ago, I offered to look after my sister Heike's two little girls, my nieces, for a few days. Or more precisely, to look after my nieces and their father, my brother-in-law, Werner. Before I began my babysitting duties, I went through Heike's extensive instructions one more time in my head: get the girls out of bed in the morning, get them washed and fed, get the older one off to preschool, and spend the day looking after the younger one; throughout the entire process try to make sure the apartment doesn't get either smashed to pieces or burned down. I prepared by rereading all my notes on paediatric emergency medicine and watching the Schwarzenegger film *Kindergarten Cop*. What could possibly go wrong?

Full of confidence, I waved my sister goodbye. But I was pushed to my parenting limits on the very first day. While I was in the kitchen trying to clear up a million razor-sharp shards of glass from the jar of gherkins I dropped and smashed on the floor — and burning lunch in the process — I suddenly heard vomiting and crying in the living room. But that was all basically manageable. After I had more or less mastered my great weakness, a nappy change, I took little Katarina along to fetch her big sister, Sophie, from preschool, and off we went to lunch. McDonald's rather than organic vegetables, thanks to my lack of multi-tasking skills.

At the end of the day, we headed off to meet Werner from

work. Arriving home, he locked the car remotely while all four doors were still open. Bang, the self-closing door on the driver's side closed automatically. Bang, bang, the passenger-side door and one of the rear doors closed. Bang, the sound of the final door closing was immediately followed by a blood-curdling scream. I ran round the car, to find Sophie with her little finger trapped in the car door. I tried frantically to open the door. 'OPEN IT!' I snapped at Werner. But as bad luck would have it, the batteries in the remote control for the central-locking system had run out, and the emergency key was wrapped in several layers of surgical tape to keep it in its plastic casing.

After we finally managed to open the door, we examined Sophie's injured finger under the kitchen lamp. Sophie was weeping bitterly. A blood blister had started forming under her nail, and her finger was blue all over. The only way to find out if the bone was damaged was to get it X-rayed. Making sure her finger was kept cool, we decided that I should drive Sophie to hospital while Werner stayed at home to look after Katarina.

Once in the car, Sophie calmed down a little, only sobbing every now and then.

'Have you ever been to a hospital?' I asked, as I started the engine.

'Yes, but I've never had anything this bad!' She was frightened, it was clear to see. 'I don't want to go there!' A big, fat tear rolled down her cheek. Seeing my niece cry so bitterly felt as if I had trapped my own finger in the car door.

'I'll be with you, so there's no need to be frightened. I know all about hospitals and I'll stay with you the whole time. I promise!'

Sophie nodded uncertainly.

I asked her, 'What turns from green to red at the flick of a switch?'

She pondered briefly. 'Hmmm. I dunno ...'

'A frog in a blender!' We both laughed.

But suddenly, her face changed and she said her finger was hurting badly. I could hardly bear to fob her off with empty words, so I switched on the hazard lights and pulled over. I found what I was looking for in the pocket of the jacket I keep in the car boot. Getting back in the car, I handed her a little white pill. 'This will make you feel better.' Less than a minute later, I asked her if the pill was working, and she nodded.

We drove over a pile of dog dirt on the side of the road, and I took it as my cue to strike up a silly song. 'Dog poo on the car tyres, squish-squish-squash, I just hope it will come off, when we give the car a wash!'

This made her roar with laughter all the way to the carpark next to the hospital's emergency wing. She endured the X-ray without complaining once. Even when the radiology assistant splayed her fingers to get a better view from the side.

'Everything okay?' I asked her.

'The tablet helped,' she said, nodding.

The radiology assistant gave me an accusatory look, as if

to say, 'You shouldn't be giving painkillers to children!'

After waiting for a while in the children's play corner, we were called in to see a nice young doctor. 'Shall we have a look at the pictures?' he asked. The problem was clear to see on the X-rays. The doctor turned to Sophie. 'It's broken,' he said with a smile.

'Really?' she answered in surprise. Then she beamed. 'Coooool!'

In the treatment room, the doctor lanced the blood blister under her nail. He was extremely careful. 'If it hurts, tell me straightaway.'

'It doesn't hurt,' said Sophie, unfazed, and gave me a smile as the doctor carefully drilled through her fingernail without any anaesthetic. 'I swallowed a painkiller!'

Finally, a nurse stuck a plaster over the nail and placed a stabilising plastic splint over her finger. Ordeal over.

Why am I telling you this story? Well, for one thing, to show off about having the coolest nieces on the entire planet. But also to show how effective it can be when you give patients the right emotional support, distract them, and give them the feeling that they're in safe hands. Oh yes ... and give them Tic Tacs. These are particularly helpful for poorly little fingers.

The well-known phenomenon that I made use of with Sophie is the opposite of the nocebo effect: the placebo effect. 'Placebo' is also a Latin word and means literally 'I shall please'. This is the phenomenon responsible for the fact that medicines that a patient believes will be effective

266

make patients feel better even if they don't contain any active ingredients. In the same way as the nocebo effect can cause a real deterioration in the patient's condition. Neither of these effects are imagined.

One kind of treatment that makes use of the placebo effect is homeopathy. It uses active agents that are sometimes so diluted that they are no longer traceable, which means that no more than a few molecules are present in the preparation. According to all scientific findings, they can't have a medical effect. If I bury an aspirin in my garden, drinking water from my well will not help cure a headache. Despite this, many people swear by the curative powers of homeopathy. And there is absolutely nothing wrong with that, as long as it doesn't go too far. If someone has a complaint and taking homeopathic globules makes them feel better, everything is fine. However, if the complaint is serious or even life-threatening, like a serious infection, or if symptoms persist, then conventional medicine is the safer option.

A large-scale meta-analysis published in the medical journal *The Lancet* came to the crushing conclusion that the physiological effects of homeopathic preparations — that is, the effects they have on our body's biological processes — are zero. The popular press reacted to this news with headlines like 'Homeopathy is all in the mind'. But that is far from the truth. Homeopathy is a very useful tool in exploiting the human body's ability to heal itself. The packaging of the active ingredient is more important than the contents in this case, but that doesn't alter the fact that patients who believe in the effectiveness of these preparations really feel much better when they take them.

Like Sophie after I'd given her a perfectly ineffectual Tic Tac, telling her it was a painkiller. But placebos don't work if they're used too often. Parents who give their children arnica globules for every little bruise sustained in the playground aren't being very clever. Bruises heal all by themselves, without medication, either homeopathic or conventional.

Giving children supposedly harmless little globules at every turn can be a dangerous mistake. From a child's point of view, everything is 'medicine', irrespective of whether it's pills, drops, or harmless globules. If they're given them too often, they will gain the impression that it's necessary to take medication for every little ache or pain. This can lead to chronic medication abuse or even full-blown addiction when they grow up.

Adults should always be free to choose the medical treatment they prefer; but for children, I would advise caution in giving them any treatment that contains, or appears to contain, active ingredients. This is because the last thing a child's heart needs is to grow up with a fondness for medication and a predisposition towards painkiller abuse.

Many over-the-counter painkillers can cause damage to the heart if taken over a long period of time. A study carried out by the Institute for Social and Preventive Medicine found, for instance, that diclofenac — a widely used painkiller — increases the risk of dying of cardiovascular causes by four times.

When my nieces fall over or graze their knees, I don't give them any 'medicine'. Those two little girls are not only funny and adventuresome, but also made of tougher stuff than you might think from looking at them. When they hurt themselves, it's enough just to hold their hands and listen attentively to them. And kiss their injuries better, of course.

The Holey Heart

'People are basically healthy!' This is one of my favourite sayings, and it's particularly true of children. As long, that is, as the heart is kept healthy with a sensible diet, enough physical exercise, and sufficient relaxation. All this is also true during the nine months a baby spends growing in its mother's belly. However, things can go wrong during that development phase in the womb. To take three examples, if an expectant mother has diabetes during her pregnancy, if she contracts German measles (rubella), or if she drinks alcohol, the baby's risk of being born with a heart defect increases. Around 160,000 babies are born with such a congenital heart defect per year worldwide, and it's usually diagnosed before the child is delivered. Sometimes, a congenital heart defect improves without treatment, but it can also represent a threat to the life of a newborn baby. With the appropriate treatment, however, nine out of ten children born with a congenital heart defect now reach adulthood.

There are many things that can go wrong as a baby's heart develops in the womb, from an incomplete development of the cardiac septa (the walls separating the chambers of the heart) to blood vessels following strange, erratic courses. Between 40 and 60 per cent of babies with the genetic defect trisomy 21,* better known as Down syndrome, are born with a congenital heart defect, most commonly a hole in the wall dividing the atria and the ventricles.† But the valves, muscles, and connective tissues of the heart can also be affected.

The most common form of congenital heart defect is a ventricular septal defect, in which there is one or several holes in the wall dividing the left and right ventricles of the heart. If the holes are small, they usually cause no outward symptoms, but if the leak between the ventricles is bigger, the effect is that every time the heart contracts, blood is pressed from the left ventricle into the right one, rather than into the aorta and out into the rest of the body. The increased pressure this causes in the right ventricle propagates into the pulmonary artery and then into the lungs, which causes increasing damage to those organs. Over time, this leads to left-ventricular heart failure.

The second-most common type of heart defect is called a tetralogy of Fallot. It involves four abnormalities of the heart: first, a ventricular septal defect as described above; second, a pulmonic stenosis, which is an obstruction of flow from the right ventricle of the heart to the pulmonary artery; third, a right ventricular hypertrophy, which is an increase in the size of the muscles of the right ventricle; and fourth, an overriding aorta. This is when the aorta is displaced so far to the right that it takes in not only oxygenated blood from

---

* These children are born with three copies of chromosome 21, rather than the normal two.

† Known medically as an atrioventricular septal defect.

the left ventricle, but also deoxygenated blood from the right ventricle, too.

Position three in the congenital heart-defect charts is an atrial septal defect, which is a hole in the wall dividing the two atria of the heart. This can cause cardiac arrhythmia and a pale blue tinge to the skin, and is usually associated with reduced physical capacity and extreme shortness of breath during physical exertion.

The blood supply in the body of an unborn child differs significantly from that of an adult. For instance, there's a connection from the pulmonary artery (the artery exiting the right ventricle of the heart) to the aorta (the artery exiting the left ventricle). This connection is called the ductus arteriosus or ductus Botalli, and performs a similar role to the foramen ovale* in bypassing the pulmonary circulation system. The foetus receives all the oxygen-rich blood it requires for its developing body directly from its mother via the umbilical cord.

The ductus arteriosus normally closes at birth. Normally, but not always. This closure actually fails so often that such a 'patent ductus arteriosus' is the fourth-most common kind of congenital heart defect, and is particularly serious for babies born prematurely. Blood now pumps from the aorta directly into the right ventricle. This often leads to too much blood in the pulmonary circulatory system, causing tiny ruptures in the blood vessels of the heart and lungs over time, which in turn can lead to heart failure and an insufficient blood

---

* Remember that? See p. 14 and p. 15.

supply to the peripheral regions of the body. The overall result is that children with this condition aren't as robust as their peers, often have cold arms and legs, and have an unusually fast and strong heartbeat (because the heart automatically tries to compensate for the insufficient supply of oxygen to the organs by working harder).

Other types of congenital heart defect include valve defects, in which one of the valves is too narrow or unable to close properly. A child's risk of being born with such a defect increases statistically with the number of family members also born with it. This clearly indicates that such children have a genetic disposition for such a defect.

With the exception of a patent ductus arteriosus, which can sometimes be closed using medication, almost all of the congenital heart defects described here require surgery to correct them. Luckily, paediatric heart surgery is now so advanced that often the only reminder of the condition in later life is a little scar. And even if all the operative and medical treatment that is available still leaves a child with a weaker than normal heart, such people can often still live long and fulfilled lives, as long as they take care not to overtax their weakened tickers.

Conclusion

Our heart is much more than just an engine. For centuries, it has been a symbol of love, lust, and passion, shot through with arrows from Cupid's bow. Although it is one of the most studied organs of the human body, there are still many questions to be answered about the way the heart, body, and mind interact with each other.

All round the world, scientists are working apace to discover the many secrets of this mysterious bundle of energy and to better understand its mechanisms, especially at the molecular level. Without modern heart research, medical progress would never have come as far as it has today, and our life expectancy would certainly not be as high as it is. Medical research stretches its feelers out into so many subjects: sleep, sex, diet — the list might go on forever. The main aim of heart research is to improve, extend, and even save the lives of patients. As I've shown, researchers have made great progress towards answering the question of how we can die of a broken heart. Broken-heart syndrome has gained increasing attention in recent years and is now one of the hottest subjects of heart research. The better this condition is understood, the better it can be treated.

However, research has already taught us one fundamental thing: a healthy heart requires a healthy body and a healthy mind. Only then can it function perfectly. Without the essential support of our other organs, it can't do its job at all. The heart is a team player. The kidneys, for example, play an essential role in regulating the amount of fluid in our vascular system and raising or lowering our blood pressure as necessary. They're aided in this by various substances produced by many different organs and tissues. These substances dilate or constrict our blood vessels as necessary, raise or lower our heart rate, or influence the strength of its beats. If all this were not the case, our heart would soon wear out.

Without all the other organs and little helpers in our blood, the heart would be little more than a single, simple cog, to use a technological analogy. The varied support it receives makes it much more than that. It is the central driving force of a highly complex mechanism. It may need to be oiled every now and again, and occasionally a part might even need to be replaced, but that happens astoundingly rarely. And the deeper our research delves into this complex clockwork, the more it becomes clear that there is no single, complete truth about the heart. We keep on uncovering more and more tiles of this intricate and artistic mosaic, only to discover that it is even more sophisticated and wide-ranging than we ever thought possible.

Even if researchers may never reach the end of this Sisyphean task, their work continues and will continue to reveal new insights, leading to ever more effective treatments for heart patients to improve and extend their lives.

Fortunately, our heart doesn't care whether we understand its workings or not. It is always there to support us.

I recently read the following quote graffitied on a wall: 'Never worry. Our hearts are like machetes; we will use them to clear a way through the jungle.'

I think this is a great tribute to a faithful friend, a powerful, enduring companion who expects nothing more in return for its selfless service than to be treated kindly.

Afterword

If you have any questions, if you are confused about anything you have read in this book, or if you believe I have omitted important points, then simply drop me a line by email, at info@herzrasenmaeher.de. I look forward to hearing from you!

Acknowledgements

There are many people without whom this book would not exist.

I would like to express special thanks to my editor, Marieke. If you had never approached me after I appeared on stage in Berlin, this project would never even have occurred to me. I thank you very much for that, and for having the patience of a saint, for all your help and all the heart you put into this project. It was a privilege to work with you on this book and a memory I will treasure forever.

I also owe a vote of thanks to a great teacher and friend, Thomas Sonnenberg, a doctor from Kiel. You are a great support in all aspects of my life. Thank you for the stimulating conversations and for your critical revision of my manuscript.

I would also like to thank all those who put me up while I was writing and/or provided so much variety and stimulation. Most of all, I would like to thank my parents and my family.

Furthermore, I also thank Simon Z., Claudia, Dirk, Zemmi, Miri, the Enzlers, in particular Bella and Christoph, Jonas, Philipp E., Heike, Katarina, Werner, Gregor, Miriam, Michael and Simon H., and the Falbs, in particular Alex, Britta, and Felix.

Many thanks also go to the Schieffer working group at the University of Marburg, the Ullstein Publishing Group, scienceslam.net (Gregor and policult), Luups, HALternativ e.V. (Tobias), scienceslam.de (Julia), the science slammers Reinhard R. and Tim G., the German–Russian Forum (Sibylle and Sandra), and all the aid organisations, emergency services, and medical wards I have had the privilege to work in.

Finally, I would like to thank Christine. It was you who put me on that stage for the first time. Thank you for the wonderful evenings and the huge amount of fun. You helped me discover something in me that I never knew existed.

Have I forgotten somebody? If so, please add your name here:

Many thanks to _____!

If it weren't for every one of you, this book would not be the way it is, or would not exist at all. Thank you very much. You all have a special place in my heart!

The key to living a happier, healthier life is inside us

Our gut is almost as important to us as our brain or our heart, yet we know very little about how it works. In *Gut*, Giulia Enders shows that rather than the utilitarian and — let's be honest — somewhat embarrassing body part we imagine it to be, it is one of the most complex, important, and even miraculous parts of our anatomy. And scientists are only just discovering quite how much it has to offer; new research shows that gut bacteria can play a role in everything from obesity and allergies to Alzheimer's.

Beginning with the personal experience of illness that inspired her research, and going on to explain everything from the basics of nutrient absorption to the latest science linking bowel bacteria with depression, Enders has written an entertaining, informative health handbook. *Gut* definitely shows that we can all benefit from getting to know the wondrous world of our inner workings.

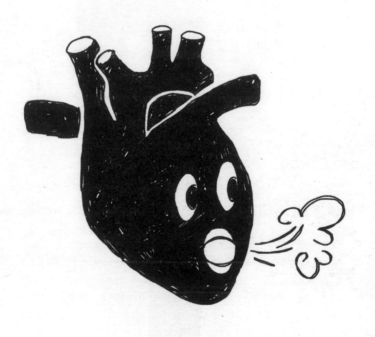